GRAPH-BASED
KEYWORD SPOTTING

SERIES IN MACHINE PERCEPTION AND ARTIFICIAL INTELLIGENCE*

ISSN: 1793-0839

Editors: **H. Bunke** (University of Bern, Switzerland)
Cheng-Lin Liu (Chinese Academy of Sciences, China)

This book series addresses all aspects of machine perception and artificial intelligence. Of particular interest are the areas of pattern recognition, image processing, computer vision, natural language understanding, speech processing, neural computing, machine learning, hardware architectures, software tools, and others. The series includes publications of various types, for example, textbooks, monographs, edited volumes, conference and workshop proceedings, PhD theses with significant impact, and special issues of the International Journal of Pattern Recognition and Artificial Intelligence.

Published

*The complete list of the published volumes in the series can be found at
http://www.worldscientific.com/series/smpai

Series in Machine Perception and Artificial Intelligence – Vol. 86

GRAPH-BASED KEYWORD SPOTTING

Michael Stauffer

University of Applied Sciences and Arts
Northwestern Switzerland, Switzerland

Andreas Fischer

University of Fribourg, Switzerland

Kaspar Riesen

University of Applied Sciences and Arts
Northwestern Switzerland, Switzerland

World Scientific

NEW JERSEY · LONDON · SINGAPORE · BEIJING · SHANGHAI · HONG KONG · TAIPEI · CHENNAI

Published by

World Scientific Publishing Co. Pte. Ltd.

5 Toh Tuck Link, Singapore 596224

USA office: 27 Warren Street, Suite 401-402, Hackensack, NJ 07601

UK office: 57 Shelton Street, Covent Garden, London WC2H 9HE

British Library Cataloguing-in-Publication Data
A catalogue record for this book is available from the British Library.

Series in Machine Perception and Artificial Intelligence — Vol. 86
GRAPH-BASED KEYWORD SPOTTING

ISBN 978-981-120-662-7

For any available supplementary material, please visit
https://www.worldscientific.com/worldscibooks/10.1142/11452#t=suppl

Printed in Singapore

I am deeply grateful to Prof. Dr. Kaspar Riesen, who gave me this unique research opportunity. In fact, his methodological, pedagogical and scientific way of working is not only very inspirational but was also a main driver for me to pursue a scientific career path.

At this point I would also like to thank Prof. Dr. Andreas Fischer for his excellent work. His profound knowledge and expertise in the field of handwritten historical document analysis and recognition has been a tremendous support and pillar for the realisation of this research project.

Special thanks go to Prof. Dr. em. Horst Bunke, who was not only a tremendous help in the publication of this book but also an inspirational pioneer in the field of graph-based pattern recognition.

I would like to thank the University of Applied Sciences and Arts Northwestern Switzerland and in particular Prof. Dr. Rolf Dornberger, Prof. Dr. Thomas Hanne, and Margaret Oertig for their support and endorsement of this book.

Furthermore, I would like to thank the Hasler Foundation Switzerland, who provided the necessary financial aid to pursue this research project.

This book is dedicated to my friends, my mother Marianne, my brother Dennis, and my partner Anniina. I will always be deeply grateful for your patience, belief, and encouragement.

Preface

Handwritten historical documents around the world are endangered by an increasing state of degradation, and thus, many libraries have started to preserve their historical treasures by digital means in recent decades. As a result, large numbers of handwritten historical manuscripts have been made available digitally for a broader audience. However, we observe a gap between the availability and accessibility of such documents. That is, automatic full transcriptions of ancient manuscripts are often not feasible due to wide variations in handwriting and noisy documents. *Keyword spotting* (KWS) has been proposed as a flexible and more error-tolerant alternative to full transcriptions. Basically, KWS allows us to retrieve arbitrary query words in handwritten historical documents.

In most cases, these methods are based on a statistical representation of handwriting images. That is, certain characteristics of handwriting images are represented by means of feature vectors. In contrast to this, few approaches can be found where handwriting is represented by means of structural representations (i.e. strings, trees, or graphs). This book explores the possibilities and limitations of graph-based representations for KWS in handwritten historical manuscripts. In particular, it introduces and thoroughly researches a novel graph-based KWS framework. First, handwritten historical document images are preprocessed and segmented into single word images. Based on preprocessed word images, graphs are extracted by means of different graph representations. The actual keyword spotting is then based on a pairwise matching of a query graph with all document graphs. The resulting graph dissimilarities are used to form a retrieval index that in the best possible case consists of all n instances of the query word as its top-n results.

Representing handwriting with graphs means that the graphs are affected by subtle variations, and thus, only inexact graph matching can be employed. We make use of *graph edit distance* (GED) as the basic matching paradigm. Basically, GED measures the minimum number of distortions needed to transform one graph into another. However, the computation of GED is exponential with respect to the number of nodes in the involved graphs, and thus, several fast but suboptimal algorithms have been proposed in recent years.

We evaluate several suboptimal algorithms for GED in the context of our novel KWS framework. We started our research with a cubic time approximation for GED. However, in cases of large documents or graphs, this algorithm has turned out to be too slow overall. Hence, we propose three different generic speed-up approaches. First, graphs are segmented into smaller subgraphs, and thus, graph matchings are conducted on smaller subgraphs rather than complete graphs. Second, we introduce a novel linear time graph dissimilarity measure in order to filter large amounts of irrelevant graph matchings. Third, we investigate recent quadratic time approximations of GED in our framework. Finally, we propose different ensemble methods that allow us to combine and use several graph representations at a time.

In an exhaustive experimental evaluation of four different historical documents, we compare the proposed graph-based KWS framework with state-of-the-art approaches for both template-based and learning-based approaches. Compared with the template-based reference systems, we observe a clear improvement in the proposed graph-based framework with respect to both accuracy and runtime. In the case of learning-based reference systems, we observe certain advantages of the reference systems in manuscripts with wide variations. However, our novel graph-based ensemble for KWS is able to reduce the influence of such variations, and thus, we observe that our framework is able to keep up with or even outperform advanced learning-based approaches.

This is particularly interesting as our novel graph-based KWS framework does not depend on any a priori learning step. The acquisition of training data is in fact an expensive and labour-intensive task in the case of handwritten historical documents. In summary, we conclude that the proposed KWS framework offers a versatile alternative to existing template- and learning-based approaches.

Michael Stauffer

Contents

List of Figures

List of Tables

Chapter 1

Introduction

1.1 Keyword Spotting (KWS) as a Scientific Discipline

Pattern recognition (PR) as a scientific discipline in computer science addresses the recognition of patterns and the correct anticipation of actions [1,2]. Basically, PR mimics a task for which humans have been trained and optimised since the early ages of the human race. In fact, the recognition of patterns and the correct reasoning about actions are crucial tasks in our everyday life. As such, the human cortex has evolved to become a true master of pattern recognition. However, with the exponential rise of *machine learning* (ML) and *artificial intelligence* (AI) techniques we have now reached a point where pattern recognition algorithms achieve better results than human experts [3–5].

Handwriting recognition (HWR) is a further application in PR, which has, in fact, a long tradition [6–9]. In contrast to machine printed documents, a difficulty of HWR is certainly large variations with respect to the style and size of handwriting images. Moreover, handwriting recognition and in particular the correct anticipation of ambiguities is highly dependent on linguistic and contextual knowledge. Hence, HWR is still regarded as a challenging task, whereas the automatic recognition of machine-printed text in documents by means of *optical character recognition* (OCR) is widely accepted as solved in the case of high-quality scans of modern prints [10].

Roughly speaking, HWR approaches by means of OCR are either based on *online* or *offline* data. In online handwriting recognition, temporal information about the writing process is available, captured by a specific electronic input device (e.g. an electronic pen or a touch-sensitive screen). In contrast to this, offline handwriting recognition is based on spatial

1

information of the scanned documents only. That is, offline handwriting recognition is generally regarded as the more difficult task [11]. In recent decades automatic handwriting recognition has been applied in many different domains such as the recognition of addresses [12–14], bank checks [15,16], handwritten mathematical expressions [17–19], whiteboard notes [20–22], or music notes [23–25], to name just a few applications.

HWR has also found widespread application in handwritten historical documents as illustrated in Fig. 1.1 [26–28]. Handwritten historical documents are an eminent witness of invaluable importance as they document both *tangible* and *intangible* aspects of historical, philosophical, scientific, and medical knowledge and events [29]. However, these important sources of knowledge are endangered by the increasing effects of degradation caused, for example, by their exposure to light and humidity, as well as the natural dissolution of paper. Hence, many ancient manuscripts have been digitally preserved by different museums, foundations, libraries, and other institutions in recent years. Manuscripts that have been digitally preserved are, for example, the Barcelona marriage database [30], the Saint Gall manuscripts [31], or the Parzival manuscript [32], as shown in Fig 1.2. Due to this global digitisation effort, ancient historical documents have not only been preserved from further degradation but have also been made available to a wider readership.

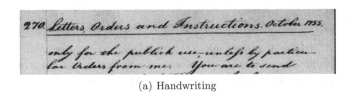

(a) Handwriting

270. Letters, Orders and Instructions. October 1755.
only for the publick use-unless by particular Orders from me. You are to send

(b) Transcription

Fig. 1.1: Exemplary handwriting recognition of a handwritten historical document.

To increase both the availability and accessibility of digitised historical documents, several international initiatives have been conducted in recent decades. In the following we provide a non-exhaustive overview

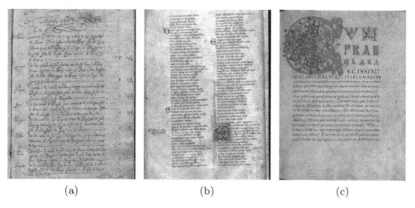

(a)　　　　　　　　　(b)　　　　　　　　　(c)

Fig. 1.2: Examples of digitally conserved historical manuscripts, namely (a) Barcelona marriage database, (b) Saint Gall manuscripts, and (c) Parzival manuscript.

over some of these initiatives [29]. In the European project *Digital Access to Books of the Renaissance* (DEBORA) [33], several Italian, French, and Portuguese books from the 16th century have been digitised and made publicly available. The European Commission-funded project *Improving Access To Text* (IMPACT) [34] aimed at an automatic and large-scale digitisation process of historical printed documents. Likewise, the initiatives *tranScriptorium* [35] and the follow-up project *Recognition and Enrichment of Archival Documents* (READ) [36, 37] have as their aim an end-to-end digitisation online platform for the transcription of handwritten historical manuscripts. Another initiative is the Catalan project *Five Centuries of Marriages* (5CofM) [30], which has made large sets of demographic records (i.e. marriage license registry entries) publicly available. Finally, the Swiss initiatives *Historical Document Analysis, Recognition and Retrieval* (HisDoc) [38], HisDoc 2.0 [39], and the ongoing HisDoc III have as their aim a large-scale end-to-end platform for the computer-aided palaeography, retrieval, and recognition of handwritten historical documents.

We conclude that both the availability and accessibility of handwritten historical documents has been largely improved through these initiatives in recent years. However, we still observe a lack of accessibility of these documents. That is, automatic and full transcription of such documents is unlikely to produce perfect results [40–42]. Reasons for this observation are manifold. First, human handwriting is inherently characterised by large inter- and/or intrapersonal variations. Moreover, handwritten historical documents are often affected by noise such as, for example, ink bleed-through and fading, degradation, or even holes in the documents as illustrated in Fig. 1.3. In addition, imperfect scanning might also negatively affect the automatic recognition of the content of such documents. Finally, the amount of training data on which HWR systems can be modelled, is often inherently limited in the case of historical documents. For these reasons, full-text transcriptions of such documents by means of HWR systems are often not feasible or are imperfect.

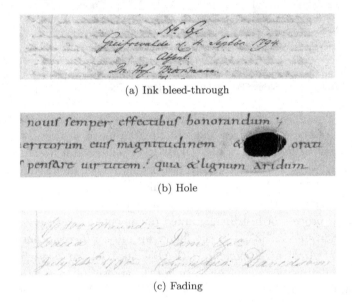

(a) Ink bleed-through

(b) Hole

(c) Fading

Fig. 1.3: Examples of variations and noise in handwritten historical documents.

To overcome these limitations, KWS has been proposed as a flexible and more error-tolerant alternative to full transcriptions of handwritten historical documents [43–45]. KWS refers to the task of retrieving all instances of a given query word in a particular document, as schematically illustrated in Fig. 1.4. Originally, KWS was proposed for speech documents [43]. Later KWS was adapted for printed [44] and eventually for handwritten documents [45]. Clearly, KWS techniques offer a convenient and powerful approach to making handwritten historical documents accessible for searching and browsing [46]. In the case of handwritten historical documents, KWS is inherently an *offline* task, and as such, limited only to spatial information about handwriting images. Offline KWS is regarded as the more complex task when compared to *online* KWS, where temporal information on the writing process is also available.

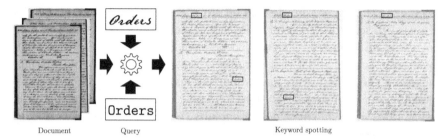

Document   Query   Keyword spotting

Fig. 1.4: Exemplary retrieval of the query "Orders". Note that found instances in the document pages are highlighted.

Most of the KWS methodologies available are based on *template-based* or *learning-based* algorithms (similar to the corresponding subfields in HWR). Early approaches of template-based KWS are based on pixel-by-pixel matchings of word images [45] by either Euclidean distance measures or affine transformations by the *Scott and Longuet-Higgins* algorithm [47]. More elaborated and error-tolerant approaches to template-based KWS are based on the matching of feature vectors that describe particular characteristics of the word images numerically, such as projection profiles [46,48,49], gradients [48], contours [50], or geometrical characteristics [51]. Also, more generic image feature descriptors have been used, such as *histograms of oriented gradients* (HoG) [52–54], *local binary patterns* (LBP) [54,55], or *deep learning* features [56], to name just a few examples. Regardless of the features currently employed, *dynamic time warping* (DTW) is the most frequently used algorithm for matching two templates, represented with sequences of features, in KWS [49–53,56,57].

Learning-based KWS is based on statistical models that have to be trained *a priori* using a (relatively large) training set of word or character images. Many approaches of learning-based KWS are based on the *hidden Markov model* (HMM) [11,32,58–63]. Early approaches are based on a *generalised hidden Markov model* (gHMM) that is trained on character images, i.e. images of Latin [11] or Arabic [59] characters. However, character-based approaches are negatively affected by an error-prone segmentation step [32]. More elaborated approaches rely on feature vectors of word images [58], for example, by means of *continuous-HMM* [61] or *semi-continuous-HMM* [61], i.e. HMMs with a shared set of *Gaussian mixture models* (GMMs). Furthermore, the use of a *Fisher kernel* has been employed in conjunction with HMMs in [60], while a line-based and lexicon-free HMM-approach is proposed in [32]. Recently, we have observed a clear shift towards *convolutional neural network* (CNN) in the field of KWS [64–68]. In most cases, CNNs are used to learn a particular word string embedding such as a *pyramid histogram of characters* (PHOC) [65–68] or a *discrete cosine transform of words* (DCToW) [66–68] that permits the retrieval of visual and textual queries in the same feature space. Further learning-based approaches are based on *recurrent neural network* (RNN) [41,57], *support vector machine* (SVM) [69–71], or *latent semantic analysis* (LSA) [72–74], to name just a few examples.

In Chapter 2 of this book, a detailed and thorough review of KWS approaches is presented.

1.2   Limitations

The focus of this book is on historical documents, and thus, offline KWS, referred to as KWS for the remainder of this book, which can only be applied. That is, any explanation, argumentation, generalisation and conclusion with respect to KWS in this book is limited to handwritten historical documents.

Moreover, the experimental evaluation of the researched graph-based KWS framework proposed in this book is limited to Latin documents, and thus, the proposed graph-based KWS framework has not been tested on non-Latin manuscripts. Note, however, that the generic graph representations would technically allow us to represent word images stemming from arbitrary languages and alphabets, and thus, the methods presented in this book could theoretically be adapted to non-Latin documents.

1.3   Research Questions

A large variety of algorithms has been developed for KWS during the last two decades [32,45,46,61,65]. However, graph-based representations and graph matching techniques have rarely been used for this specific task [75–77]. This is rather surprising as graphs are inherently capable of representing subtle variations in handwriting. That is, graphs — in contrast to feature vectors — are able to adapt their size and structure to the size and complexity of an underlying pattern. Moreover, graphs are able to represent binary relationships between substructures of a pattern. The limited usage of graphs in KWS is probably due to well-known problems arising with graphs in the field of unconstrained handwriting recognition. However, KWS is not necessarily based on handwriting recognition. In fact, it turns out that the paradigm of graph matching is able to meet the overall requirements of KWS. That is, through the representation of isolated words with graphs, the search and retrieval in documents can be regarded as matching a query graph (keyword) with a large set of document graphs.

The major objective of this book is to employ various graph-based pattern recognition techniques within the field of keyword spotting in handwritten historical documents. That is, the main focus of this book is to obtain an in-depth understanding of the advantages and limitations of graph-based methods in the field of KWS.

As this is one of the first times that graph-based methods have been employed for the task of keyword spotting, a number of open research questions have to be addressed. Clearly, the overall question to be answered is whether graph-based representation and especially graph matching techniques can be usefully employed for this task. In the following paragraphs, the two major lines of research pursued in this book and the respective open research questions are described in detail.

The first issue to be solved in this book is that of extracting graphs from the underlying documents. Basically, there are various possible ways of representing handwriting by means of graphs. One could, for example, use several handwriting characteristics given by keypoints or projection profiles as a basis for graph representations. In fact, it turns out that graphs are well-suited to the formal representation of such information, as graphs are able to represent structural relationships between subparts of this particular information. Moreover, graphs are able to adapt their size and structure to the complexity of the underlying handwriting. However, which actual graph representations will be best-suited for the task of graph-based HWR and, in particular, graph-based KWS is an open research question.

Given that the handwritten words in the underlying documents, as well as the keyword itself, are represented by means of graphs, the second issue is to develop a graph matching framework for the task of graph-based KWS. Due to the variability in handwriting with respect to style and scaling, a flexible graph matching model is needed. The major drawback of many graph matching algorithms is their computational complexity that restricts their applicability to graphs of rather small size. However, in the last decade, powerful approximation algorithms have been proposed for the computation of graph dissimilarities (e.g. [78,79]).

For this book, we adapt and substantially extend several existing frameworks for approximate graph distance computation to the problem of matching handwriting word graphs. Beside the adaptation of some general graph matching frameworks to a novel problem specification, we also aim to develop a completely novel graph matching framework that better considers the particular characteristics of handwritten graphs as well as KWS in general. In summary, KWS does not only require a comprehensive graph representation but also large amounts of graph matchings, depending on the size of the manuscript.

To test the researched methods, four handwritten historical documents will be used, namely *George Washington* (GW) [32], *Parzival* (PAR) [32], *Alvermann Konzilsprotokolle* (AK) [80], and *Botany* (BOT) [80]. All of these historical documents are well known in the community of document analysis, and moreover, benchmark tests for KWS are available on them. That is, our overall goal is to develop a novel graph-based KWS framework that is able to keep up, or even outperform, template-based and learning-based state-of-the-art approaches. In particular, we aim to evaluate whether or not the proposed KWS approach is able to keep up with respect to runtime (measured by the average matching time per pair of graphs or sequence of feature vectors) and accuracy (measured by the *mean average precision* (MAP) in the case of local thresholds and *average precision* (AP) in the case of global thresholds).

1.4   Contributions

The research outcomes of this book have been presented over the last three years, in one book chapter [81], three international journal papers [82–84], and eight papers in the proceedings of international workshops and conferences [85–92]. The contribution is summarised in the following list.

- Four novel graph formalisms for the representation of handwritten word images are introduced in [86]. An initial approach is based on the representation of characteristic points by nodes, while edges represent strokes between these points. A further approach is based on a grid-wise segmentation of word images, where each segment is eventually represented by a node. Finally, two representation formalisms are based on vertical and horizontal segmentations of word images by means of projection profiles. In order to provide handwriting graphs and benchmark results to the scientific research community, we launched a website focusing on the promotion of graph-based approaches for KWS in the field of handwritten historical documents (see http://www.histograph.ch/).

- We adopted the *bipartite graph edit distance* (BP) algorithm [93] to create a fully-fledged graph-based KWS framework. That is, we developed and researched novel cost models, provided exhaustive optimisations of the system, introduced various normalisation models, and provided visualisations of graph matchings [85]. The proposed KWS framework has been thoroughly evaluated with respect to KWS accuracy in two threshold scenarios for four benchmark datasets. That is, we provide a detailed comparison with four template-based and three learning-based state-of-the-art systems on all four manuscripts [81, 82].

- We researched various improvements with respect to runtime performance so that our algorithmic framework becomes applicable to large documents. To this end, we focused on various possibilities to speed up the basic graph matching process actually used for KWS. First, we propose different heuristics (in particular graph segmentations and aggregated graph matchings) as well as filtering methods to reduce the runtime of the KWS approach. In the case of segmentations, graphs are divided into smaller subgraphs, and thus, graph matchings can be carried out on smaller subgraphs rather than complete graphs [90]. Filtering methods, on the other hand, allow us to omit large amounts of irrelevant matchings without the computationally demanding graph matching algorithm [83, 88].

- Another line of research pursued to improve the runtime performance is to adopt and employ different graph matchings with quadratic rather than cubic time complexity, namely *Hausdorff edit distance* (HED) [79], *context-aware Hausdorff edit distance* (CED) [94], and the *bipartite graph edit distance 2* (BP2) algorithms [95]. Quite astonishingly, we observed not only runtime improvements when using these quadratic time algorithms but also improvements with respect to accuracy [84, 89, 92].

- Moreover, we improved the KWS accuracy in our framework in order to keep up with recent learning-based KWS systems. That is, we observed that learning-based approaches are especially beneficial in cases where documents offer large intraword variations (in style and scaling). In order to achieve comparable results for these challenging datasets, we propose approaches to combine different graph-based KWS systems by means of different ensemble methods. These methods require no learning, but are able to keep up with learning-based reference systems [91].

1.5 Outline and Organisation of the Book

This book is organised into seven chapters.

- **Chapter 2** proposes a novel taxonomy for *keyword spotting* (KWS). According to this taxonomy, KWS approaches are distinguished with respect to representation formalisms (i.e. statistical vs. structural), classification approaches (i.e. template-based vs. learning-based), and query input possibilities (i.e. *query-by-example* (QbE) vs. *query-by-string* (QbS)). Following this taxonomy, a thorough and comprehensive review of state-of-the-art KWS methods is provided.
- In **Chapter 3**, four different historical manuscripts, namely the *George Washington* (GW) letters, the *Parzival* (PAR) manuscript, and the *Alvermann Konzilsprotokolle* (AK) and the *Botany* (BOT) documents, are described. These sets of documents or manuscripts are employed in the remainder of this book for experimental evaluations. All four manuscripts are affected by particular signs of degradation, and thus, different image preprocessing techniques have to be applied. Moreover, in our particular case, document pages need to be segmented into individual word images. Both steps (preprocessing and segmentation) are thoroughly discussed in this chapter.
- **Chapter 4** proposes four different graph representations for handwritten word images. The preprocessed and segmented handwritten word images can be represented by means of any of the four formalisms, and thus, a complete document can be transformed into sets of graphs.
- In **Chapter 5** we first review different exact and inexact graph matching paradigms. To handle variations in the handwritten graph representations, we then focus on inexact graph matching and in particular on *graph edit distance* (GED). GED allows us to endow a certain error-tolerance with respect to both structure and labelling in the matching process. Hence, this paradigm is particularly well-suited to matching handwriting that often includes

subtle variations. We review four different suboptimal (but fast) algorithms for the computation of GED and introduce a novel linear time graph dissimilarity model.

- Next, in **Chapter 6**, we condense the methods of the previous chapters into a self-contained graph-based KWS framework. Moreover, we propose different extensions of the basic framework that permit improvements with respect to both runtime and accuracy.

- In **Chapter 7**, a thorough experimental evaluation of the novel approaches to all historical manuscripts is provided. First, we exhaustively compare the different graph-based KWS approaches with each other. Second, the graph-based approaches are compared with both state-of-the-art template-based and learning-based methods.

- Finally, **Chapter 8** concludes the main results of this book and provides possible lines of research that might be worth pursuing in future work.

Chapter 2

Related Work

2.1 A Taxonomy of KWS Systems

Handwritten historical documents are often affected by various variations caused, for example, by the handwriting itself (style and size), as well as signs of degradation (e.g. ink bleed-through, fading, holes, etc.). Moreover, the size of available training data is often limited in the case of ancient manuscripts. Consequently, automatic full transcriptions of such documents are often imperfect. *Keyword spotting* (KWS) has been proposed as a flexible and more error-tolerant alternative to full transcriptions (see Section 1.1). KWS permits the retrieval of arbitrary query words in handwritten historical documents. In particular, KWS frameworks are typically based on three subsequent process steps, as described in the following paragraphs and shown in Fig. 2.1.

- (A) First, document images are generally preprocessed in order to minimise variations caused, for instance, by noisy background images, skewed scanning, or degraded documents. Optionally, document images are segmented into line- or word images.
- (B) On the basis of preprocessed images $\{w_1, \ldots, w_N\}$ (i.e. handwritten documents, line or word images), specific handwriting characteristics are extracted and represented by means of a formal representation $\{R_1, \ldots, R_N\}$ (e.g. feature vectors, strings, graphs, etc.).
- (C) The set of formal representations is then used to query unknown words (i.e. document words). That is, a query q (shown as visual or textual input) is used to retrieve all individual entities of this word in $\{R_1, \ldots, R_N\}$. In the best possible case, this index

represents all n instances of a given query word observable in the document as its top-n results.

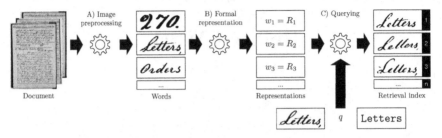

Fig. 2.1: Keyword spotting process consisting of three subsequent steps, namely image preprocessing, formal representation, and querying.

Existing KWS frameworks can be distinguished with respect to different characteristics, as shown in the taxonomy in Fig. 2.2.

First, KWS approaches can be distinguished with respect to the formalism actually used to represent the underlying image data (level 1 in the taxonomy tree in Fig. 2.2; see also step (B) in Fig. 2.1). In the case of *statistical* representations, certain characteristics are represented by means of (sequences of) feature vectors. In the case of *structural* representations, the inherent characteristics of the handwriting are represented by means of strings, trees, or graphs (whereby strings and trees can be seen as special cases of graphs).

KWS can be further divided into *template-based* and *learning-based* approaches (level 2 in the taxonomy tree in Fig. 2.2; see also step (C) in Fig. 2.1). Template-based approaches directly retrieve a query word (for example, represented by a sequence of feature vectors) by means of pairwise matchings with a set of document word images (each represented by a sequence of feature vectors). Learning-based approaches on the other hand are based on a statistical model that has to be trained *a priori* for the actual classification task. Given an arbitrary query, the classifier should then be able to classify the document content into relevant and irrelevant words by means of the learned model.

Finally, KWS approaches can also be distinguished with respect to the type of query, namely *query-by-example* (QbE) and *query-by-string* (QbS) (level 3 in the taxonomy tree in Fig. 2.2; see also step (C) in Fig. 2.1). In the case of QbE, the user selects an exemplary word image of the keyword to be retrieved in the document, while in the case of QbS, a keyword can be retrieved by means of a textual input. In fact, a taxonomy by means of the query input possibility (i.e. QbE and QbS) is probably the most popular approach to categorise in the field of KWS [60, 66, 69, 96–98]. However, in this book we focus more on the comparison of template-based and learning-based approaches as the amount of training data is often limited in the case of handwritten historical documents.

Note that there are KWS approaches with respect to all dimensions of the taxonomy, except learning-based QbS approaches that employ structural data. This is why the right outer paths in the taxonomy of Fig. 2.2 are shown with dotted lines. Moreover, we observe that most approaches for QbE are based on template-based methods, while most QbS approaches are based on learning-based methods. That is, either of the classification approaches is inherently more suited for a specific query input possibility.

If we compare the three different features used in our taxonomy with each other, several advantages and disadvantages can be observed for one side or the other. These pros and cons are briefly described in the following three paragraphs.

Statistical vs. Structural Representations If we compare the two kinds of representation formalisms, we can observe that statistical approaches offer both a rich mathematical environment and low computational complexities for most of the basic operations needed to conduct KWS. On the other hand side, structural approaches, in particular graphs, are able to explicitly model binary relationships that might exist in the underlying patterns (which is not possible with the statistical approach). Moreover, strings, trees, and graphs are able to adapt both their structure and size to the actual size and complexity of the pattern being represented (while feature vectors have to preserve the predefined size in all cases). Thus, structural approaches result in more powerful and flexible representations when compared to those derived from statistical approaches. However, the structural domain lacks a mathematical structure and we observe a general increase in the complexity of many basic operations applied in the structural domain.

Template-based vs. Learning-based Generally, learning-based KWS methods result in a higher degree of accuracy when compared to template-based approaches. Moreover, their ability to cope with particular variations in handwriting is often higher and therefore it is often the case that fewer image preprocessing steps are necessary. However, this advantage is

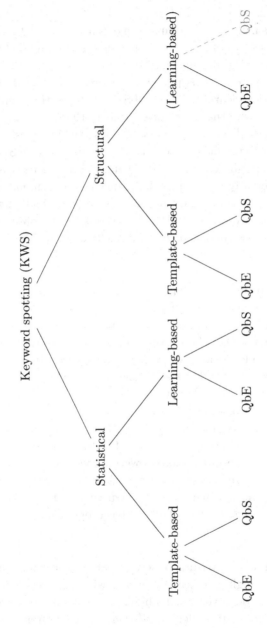

Fig. 2.2: Keyword spotting taxonomy of common approaches.

accompanied by a loss of flexibility and generalisability, which is due to the need to learn a statistical model on some training data. In fact, the accuracy of many learning-based approaches is crucially dependent on the size and quality of the labelled training data. However, the labelling of historical handwriting is often a laborious and costly process. A further limitation of this approach may be the fact that the size of the training data is often inherently limited by the size of the document (many historical documents consist of only a few pages).

QbE vs. QbS Finally, if we compare the two different query approaches with each other, QbS seems to be a more natural retrieval possibility when compared to QbE. Moreover, QbS generally allows *out-of-vocabulary* retrievals. However, these advantages are accompanied by an additional processing step, namely the generation of synthetical images for template-based approaches or the learning of a statistical model for learning-based approaches.

The remainder of this chapter reviews a wide range of related literature. This review is structured with respect to the KWS process steps (A)-(C) and the resulting KWS taxonomy shown in Figs. 2.1 and 2.2, respectively. First, we review common image preprocessing methods often used in KWS in Section 2.2. Next, different approaches for the statistical or structural representation of handwriting are reviewed in Section 2.3. Finally, these formal representations are used for querying, as reviewed in Section 2.4.

2.2 Image Preprocessing

The first step in any KWS system is often the image preprocessing of scanned handwritten document pages (see Fig. 2.3), regardless of the representation formalism and classification method employed. Image preprocessing aims to reduce variations caused by different handwriting styles (i.e. interpersonal variations) and the document itself (e.g. pixel noise, skewed scanning, or degraded documents). In the following, we focus on the most relevant image preprocessing steps, namely noise removal, binarisation, skew and slant correction, segmentation, and normalisation (with respect to both size and baseline).

In Table 2.1, the different image preprocessing steps and some seminal references are summarised. In general, there is no fixed order for the application of the different image preprocessing steps, but in most cases a certain filtering and binarisation are employed first.

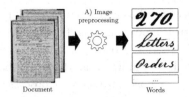

Fig. 2.3: Step (A): Image preprocessing of scanned handwritten document images.

Table 2.1: Common image preprocessing steps for keyword spotting.

Preprocessing	
Binarisation	[32, 40–42, 45, 50, 56, 57, 59, 76, 96, 99–116]
Filtering (noise removal)	[32, 40, 45, 46, 53, 55, 58, 63, 71, 76, 99, 100, 104, 106, 117–121, 121]
Skew correction	[32, 40–42, 46, 52, 56–61, 63, 96, 102, 103, 105–107, 116–119, 121–123]
Slant correction	[32, 40–42, 46, 52, 53, 56–58, 60, 61, 63, 96, 100, 102, 103, 105–107, 117, 122, 123]
Segmentation	[11, 32, 40–42, 45, 46, 50, 52–61, 63–66, 68, 69, 71, 73, 75–77, 96–111, 114–119, 121, 122, 124–135]
Size normalisation	[32, 40–42, 52, 56, 57, 60, 61, 71, 96, 100, 102, 103, 106, 107, 112, 116, 118, 119, 121, 123, 125, 132]
Baseline normalisation	[32, 40, 41, 46, 57, 58, 96, 100, 102, 103, 117, 119, 122]

In a wide range of KWS applications (e.g. [32, 45, 50, 57, 99, 100, 105]), greyscale images are first transformed into binary images in order to separate the ink from the background. That is, each pixel is assigned either to the foreground or background of the corresponding image. The decision whether a pixel is assigned to the foreground or background is often based on a global threshold (applied, for instance, in [32,45,56,57,101,109]). That is, pixels below the given threshold are regarded as background, while pixels above the threshold are regarded as foreground. Rather than manually/visually defining a threshold, one often uses adaptive thresholding by means of *Otsu's* method [136] (e.g. [97,101,137]), *Niblacks* algorithm [138], or the *Savuola* algorithm [139] (e.g. [50,116]).

As many historical documents are affected by noise (e.g. ink bleed-through, faded ink, stains, smearing, etc.), filtering such artefacts is a crucial image preprocessing task. The basic idea of filtering is to maximise both the number of true foreground pixels (i.e. ink) as well as the number of true background pixels. Often-used approaches are, for example, based on morphological operators [99, 121], *difference of Gaussians* (DoG) (applied, for instance, in [32, 40, 45]), and *Laplacian of Gaussian* [100], or contour smoothing [63, 120, 121].

Another crucial image preprocessing task is the segmentation of document images into line or word images (e.g. [32, 45, 52, 53, 58, 65, 96, 105, 108]). However, the manual segmentation of handwritten historical documents is often labour and cost intensive. Thus, many frameworks make use of a (semi-)automatic segmentation. In particular, document images are first segmented into lines of text, and then further segmented into single word images. For line segmentation it is common to use vertical projection profiles of the document image and segment at minima of the corresponding histogram. The same approach can be employed for the extraction of word images by means of horizontal projection profiles of the resulting text line images. However, the segmentation of handwritten historical documents is often far from perfect and affected by segmentation errors (i.e. under- and over-segmentation) [74].

Handwritten documents are also often affected by variations in the handwriting itself. To minimise the influence of such variations, different approaches are generally applied, such as, for example, skew and slant correction, size (height and/or width) normalisation, and baseline correction. Skew — the inclination of the document (see Fig. 2.4a) — can be caused by both the writer of the document and imperfect (skewed) scanning. In many applications (e.g. [32, 52, 56, 58, 96, 117]), the skew is estimated on lines of text by means of a linear regression of the lower baseline (e.g. [40, 58, 116, 140]), or projection profiles [141]. The estimated skew can then be removed by means of rotation. Moreover, the slant — the inclination of the characters (see Fig. 2.4b) — is often corrected in a similar way (e.g. [32, 52, 53, 58, 96, 105, 117]). In particular, the slant is often estimated by means of the mean gradient [105] or the radial contour point distribution (e.g. [32, 58, 102, 103, 140]) and then corrected by means of shear transformations.

In many handwritten historical documents the size and position of words can vary greatly even if they represent the same word class. To mitigate such variations, many of the proposed approaches normalise (i.e. scale) the size of the word or line to a predefined width and/or height (e.g. [32, 41, 52, 100, 116]). Also, the position of handwritten artefacts might not always be perfectly aligned with respect to their segmentation, often referred to as a *bounding box*. Hence, the lower baseline (see Fig. 2.4c) — the imaginary line that people write on — is often-used to vertically align the handwriting. In many applications this is done by means of horizontal projection profiles (e.g. [41, 57, 122]), and/or linear regression [41, 57, 102].

| (a) Skew angle | (b) Slant angle | (c) Upper and lower baseline |

Fig. 2.4: Characteristics of handwriting such as skew (a), slant (b), and upper and lower baseline (c).

2.3 Formal Representation

On the basis of preprocessed document, line, or word images, a formal representation (i.e. a specific data structure) needs to be defined in order to encode the images as illustrated in Fig. 2.5. These formalisms are used to represent certain characteristics of the handwriting, used to differentiate between relevant and irrelevant words in the KWS task. In general, we distinguish between *statistical* and *structural* representation formalisms. In the former case, the handwriting is represented by means of feature vectors (reviewed in Section 2.3.1), while in the latter case the handwriting is represented by means of strings, trees, or graphs (reviewed in Section 2.3.2).

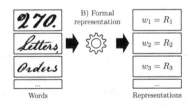

Fig. 2.5: Step (B): Formal representation of handwritten word images.

2.3.1 *Statistical Representation*

In the case of statistical representations, feature vectors are extracted from the images on either a *global* or *local* scale. In the case of global representations (also known as a *holistic approach*), features are extracted on whole handwriting images (i.e. document, line, or word images), and thus, represented by a single (or global), often high-dimensional feature vector $\mathbf{x} \in \mathbb{R}^n$. In the case of local representations, subparts of the handwriting images are independently represented by local feature vectors. As a result, a sequence of $t > 1$ feature vectors is acquired for every handwriting image i.e. $\mathbf{x}_1, \ldots, \mathbf{x}_t$ with $\mathbf{x}_i \in \mathbb{R}^n (i = 1, \ldots, t)$.

Early KWS approaches make use of global pixel-based representations of word images [45, 46, 49, 100, 117]. That is, a word image is represented by a single binary feature vector to represent the matrix of all fore- and background pixels. However, more enhanced features have also been applied in a global scenario (e.g. [48, 125, 142–144]). Regardless of the features employed, global representations are often negatively affected by noise and variations in handwriting, and thus, more recent approaches make rather use of local representations.

Local feature descriptors are used to describe aspects of handwriting images. In many cases (e.g. [32, 42, 53, 117]) these features are acquired by a sliding window approach, i.e. a window with t pixel width that moves seamlessly over the handwriting image from left to right. For every window position, a feature vector is extracted that describes the arrangement of the pixels in the current window. This procedure results in a sequence of feature vectors for every image. Local features can alternatively be extracted by grid-wise segmentations [70, 111, 120, 145], keypoint regions [111, 146], or quadtree segmentations [55].

Basically, we distinguish between three types of feature descriptors, namely *handwriting descriptors*, *texture and shape descriptors*, and *encoding and embedding approaches*. The first type of features aims to represent certain characteristics of the actual handwriting, while the second type of features is based on more generic descriptors. Finally, encoding and embedding approaches are mostly based on spatial or textual embedding of the first two types of features. In the following paragraphs, the three types of features are thoroughly reviewed. Note that most of the feature descriptors reviewed in the next sections can be applied in both global and local scenarios.

2.3.1.1 *Handwriting Descriptors*

The first type of feature descriptors is based on certain characteristics of the handwriting, as shown in Table 2.2. An example of these kinds of features is vertical and horizontal projection profiles of handwriting images (e.g. [58, 99,115,117,118,147]). That is, a histogram is used to represent the number of foreground pixels for each column and row in a vertical or horizontal direction, respectively. Likewise, word profiles recognise the first foreground pixel from the top (upper word profile) or bottom (lower word profile) of each column of a handwriting image [46,58,97,99,117,118,122]. Also the number of fore- and background transitions per column is used as a feature descriptor [46,97,99,106,117,121,122]. Quite a number of KWS approaches are based on the detection of certain points such as, for example, contour (or edge) points [45, 46, 46, 50, 100, 101, 117], corner points [49, 117], or other keypoints [130]. Predefined sets of features have also been proposed such as, for example, the so-called *Marti* [51] or *Vinciarelli* [148] features. The former set of features is based on nine different geometrical features, while the latter is based on the number of foreground pixels (termed *pixel count*) per segment in a fixed sized grid. A similar set of features, employed for KWS in [63], has been proposed in [149]. This feature set is based on 26 different density features (that are similar to the features proposed in [51]). Further handwriting descriptors are based on the greyscale variance (termed *intensity*) [122] or the centre of gravity of handwriting images [115].

Table 2.2: Handwriting descriptors of common keyword spotting approaches.

Feature descriptor	
Projection profiles	[46, 48, 49, 58, 99, 115, 117, 118, 122, 147, 150]
Word profiles	[46, 58, 97, 99, 117, 118, 122]
Background transitions	[46, 97, 99, 106, 117, 121, 122]
Contour points	[45, 46, 46, 50, 100, 101, 117]
Corner points	[49, 117]
Keypoints	[130]
Geometrical characteristics	[32, 40–42, 52, 57, 61, 96, 102, 103, 107, 115, 119, 121]
Pixel count	[52, 61, 70, 119, 121]
Others	[63, 115, 122]

2.3.1.2 *Texture and Shape Descriptors*

The second type of feature descriptors is based on more generic texture and shape descriptors that have also been applied in other pattern recognition applications, as shown in Table 2.3. The texture descriptors in a first group are based on *image kernels* (also referred to as *filters*) (e.g. [52, 71, 106, 118, 122, 127, 143]). That is, an input image is convolved by a specific kernel (i.e. a small matrix) to create an output image with a particular characteristic such as blur, edge or gradient enhancement. This output image can then be used to extract specific features (e.g. the direction of gradients). In the case of KWS, these kernels are often based on filters such as *Gabor* filters [124, 125] or *Gaussian* filters [122, 143, 144]. Functions of *Moments* [151] are also used for translation and scale invariant representations of the handwriting images [125]. Other popular feature descriptors, employed in [48, 121, 125, 142] for instance, are the so-called *gradient, structural and concavity* (GSC) features [152, 153]. These features are based on a grid-wise segmentation from which three types of features (namely gradient, structural, and concavity features) are extracted and concatenated to form a feature vector.

The most popular texture descriptor for KWS, used in [42,52–54,70,110, 112] and others, is probably *histograms of oriented gradients* (HoG) [154]. For HoG features a (sub)image is first filtered by a Gaussian kernel to detect gradients. Following this, the filtered image is segmented grid-wise. For each segment a fixed number of orientation bins is then used to describe the direction of the gradient vectors. Finally, the single bins are concatenated to form one histogram. Similar to the idea of HoG features, *scale-invariant feature transform* (SIFT) [155] is also based on the orientation of gradients. However, gradient histograms are extracted around the context of certain keypoints [62,69,70,113], rather than with a sliding window, as in the case of HoG. A similar idea is pursued by *loci* features [156] which measure the number of intersections in a predefined number of directions for every background pixel. *Local binary patterns* (LBP) [157, 158] have also been employed for KWS in recent years [54, 55, 112]. LBP describe a certain neighbourhood (for instance 16×16 pixels) by means of circular structures. A vector can then be constructed by following the circular structure in either a clock-wise or an anti-clock-wise direction. This procedure is normally repeated for different radii, and thus, one global histogram can be created by concatenating the single binary vectors.

In contrast to these manually engineered features, a number of *deep learning* features, for example, based on *convolutional neural networks* (CNNs), have been proposed in the context of KWS [42,56,98]. Rather than extracting features by means of a certain predefined procedure (e.g. GSC, SIFT, HoG, or LBP), this kind of features is learned on images in an unsupervised manner by means of CNNs.

In contrast to texture descriptors, shape descriptors are used to represent the actual two- or three-dimensional shape of an object. In the case of *shape context* [159], for instance, sparse contour points of an image are radially segmented by means of the polar-coordinate system [46, 49, 100, 117]. For each segment, the number of contour points is counted, and thus, a histogram can be created by concatenating the single segments. A similar shape descriptor is the *blurred shape model* [160] used for KWS in [110]. In the case of the *pair of adjacent segments* [161], pairs of adjacent contour structures are used as shape descriptors [109]. Finally, the *heat kernel signature* [162] is based on the concept of heat diffusion of three-dimensional objects, adapted in [146] to handwriting images in a KWS application.

Table 2.3: Texture and shape descriptors for common keyword spotting approaches.

Feature descriptor	
Texture	
- Kernels	[122, 124, 125, 143, 144]
- Moments	[125]
- Gradient, structural and concavity	[48, 121, 125, 142]
- Histograms of oriented gradients	[42, 52–54, 60, 61, 70, 71, 106, 110, 112, 115, 118, 121, 127, 132]
- Scale-invariant feature transform	[62, 69, 70, 113]
- Loci	[97, 104]
- Local binary patterns	[54, 55, 112]
- Deep learning features	[42, 56, 98]
Shape	
- Shape context	[46, 49, 100, 117]
- Blurred shape model	[110]
- Pair of adjacent segments	[109]
- Heat kernel signature	[146]

2.3.1.3 *Encoding and Embedding Approaches*

The last type of feature descriptors is based on an encoding or embedding of features, as shown in Table 2.4. In the case of encoding approaches, features are not directly used for classification but rather compared to a set of prototype features or probabilistic models. In the case of *bag-of-visual-words* (BoVW) (also known as *bag-of-features*) [163], a codebook of images (termed visual vocabulary) is first described by a particular feature descriptor and then clustered by means of k-means [164]. Each segment of an image is then assigned to its next cluster (termed *visual word*), and thus, a histogram can be created by averaging the number of visual word occurrences per image (this process is also known as *pooling*). BoVW has been used in combination with different feature descriptors such as HoG [131], SIFT (e.g. [72–74, 97, 105]), loci [111], or shape context [111].

In contrast to BoVW, *Fisher vectors* (FV) [165] make use of a flexible model, namely a *Gaussian mixture model* (GMM), for the codebook generation. That is, a sample image is described with respect to its deviation from the (fixed length) parameter of a generative model by means of a log-likelihood function. When compared to BoVW, the idea of FV is regarded as the more flexible and powerful approach due to the generative kernel method [166]. FV have been employed in conjunction with both HoG [60] and SIFT [69, 70, 128] features for KWS.

Many global feature descriptors result in high-dimensional feature vectors (e.g. BoVW with large codebooks). However, the discriminative capability of such high-dimensional feature vectors is often low due to its sparse representation. Thus, different approaches aim to reduce the high-dimensionality while preserving its global structure. In general, this reduction is either based on linear functions (e.g. *principal component analysis* [167]) or non-linear functions (e.g. *Isomap* [168], or *t-distributed stochastic neighbour embedding* (t-SNE) [169]). The low-dimensional mappings are regarded as manifold of the high-dimensional feature space [116], and thus named *manifold embedding*. In the case of KWS, both Isomap and t-SNE have been used for a non-linear dimensionality reduction of feature vectors [116, 133].

Recently, *language models* (LMs) and *word string embeddings* have been used to provide additional statistical and structural language content. Characters do not appear arbitrarily in a text but with a degree of probability that can be learned *a priori* on a given corpora of text [107]. In conjunction with extracted sequences of feature vectors of word images, the estimated textual appearance probabilities can then be used to support and improve a certain statistical model (e.g. *hidden Markov models* (HMMs)) [51, 107, 148]. In particular, the log-likelihood function in HMMs can be extended by probabilities of both features and appearance rather than features only. However, language models and word string embedding methods can also be derived from word images rather than textual corpora by means of learning-based methods [66, 73, 128]. The resulting embeddings can then be used as a kind of feature representation for KWS. In the following paragraphs, we summarise some recent approaches.

In the case of n-gram LMs, the probability of n text patterns appearing in a row is estimated [66, 73, 107]. Example n-grams for the word Keyword are: 1-grams (unigrams) = {K, e, y, w, o, r, d}, 2-grams (bigrams) = {Ke, ey, yw, wo, or, rd}, 3-grams (trigrams) = {Key, eyw, ywo, wor, ord}, etc.

The drawback of n-grams is the lack of spatial information. For instance, the unigram histogram for *silent* is the same as for *listen* [69]. *Word string embedding* approaches add an additional segmentation prior to the extraction of unigram histograms. The *pyramid histogram of characters* (PHOC) [69, 128], for instance, splits words on four different levels into two, three, four, and five segments. For each segment, a unigram histogram with all 36 Latin characters is created and then used to form one global histogram. PHOC embedding has become increasingly popular in KWS in recent years (e.g. [65, 68, 128, 170]). However, other word string embedding approaches have also been proposed, such as, for example, *discrete cosine transform of words* (DCToW) [66], *Levenshtein space deep embedding* (LSDE) [68], or *spatial pyramid of characters* (SPOC) [67]. Finally, bi-grams can also be modelled by means of *word graphs* (WGs) [123], where nodes are used to represent word boundaries, while edges are used to represent word likelihoods[1].

2.3.2   *Structural Representation*

Structural representations are based on powerful data structures such as strings, trees, or graphs. Also, graph-like representations have become more popular in recent years (see Table 2.5), for example, the *inkball* model [108, 114]. Graphs (and graph-like structures) are — in contrast to feature vectors — flexible enough to adapt their size to the size and complexity of the underlying handwriting image. Moreover, graphs are capable of representing binary relationships that might exist in different subparts of the underlying pattern. However, graph-based and graph-like KWS approaches are rather limited to date [75–77, 97, 108, 109, 114, 129, 172]. That is, the vast majority of the KWS research still makes use of statistical rather than structural representations. At first glance, this is rather surprising as the inherent properties of structural representations would be well-suited to the representation of handwriting. The main reason for sacrificing the power and flexibility of graphs is the increased complexity of many basic operations on graphs.

---

[1] The actual representation of words is still based on sequences of feature vectors, and thus, we regard this approach as a statistical, rather than structural, method.

Table 2.4: Encoding and embedding methods of common keyword spotting approaches.

Feature descriptor	
Bag-of-visual-words	
- Histograms of oriented gradients	[131]
- Scale-invariant feature transform	[62, 72–74, 97, 105, 120, 171]
- Loci	[111]
- Shape context	[111]
Fisher vectors	
- Histograms of oriented gradients	[60]
- Scale-invariant feature transform	[69, 70, 128]
Manifold embedding	
- Isomap	[133]
- t-Distributed stochastic neighbour embedding	[116]
Word string embedding / language model	
- n-grams	[66, 73, 107]
- Pyramid histogram of characters	[65, 67–69, 128, 135, 170]
- Discrete cosine transform of words	[66–68]
- Levenshtein space deep embedding	[68]
- Spatial pyramid of characters	[67]
- Word graphs	[123]

In general, a graph is defined as a four-tuple $g = (V, E, \mu, \nu)$ where V and E are finite sets of nodes and edges, and $\mu : V \to L_V$ as well as $\nu : E \to L_E$ are labelling functions for nodes and edges, respectively. In most of the graph-based KWS approaches [75, 97, 108, 114, 129, 172], certain characteristic points — so-called keypoints — in the handwriting image are represented by nodes (e.g. start, end, and intersection points of the handwritten strokes). Edges are then used to represent strokes between these keypoints. Invariants (referred to as *graphemes*), which can be seen as a codebook of strokes, are also used to extract nodes from images [76, 77]. Similarly to the keypoint representation, edges are used to represent the connectivity between the strokes.

2.4 Querying

The actual keyword spotting task is based on the matching or classification of the extracted feature vectors (in the case of statistical representation) or graphs (in the case of structural representations) by means of *template-*

Table 2.5: Structural feature descriptors of common keyword spotting approaches.

Feature descriptor	
Graphs	
- Keypoints	$[75, 97, 129, 172]$
- Invariants	$[76, 77]$
Inkballs	
- Keypoints	$[108, 114]$

based or *learning-based* algorithms, as shown in Fig 2.6. That is, a query word q (used to represent a certain keyword) is searched for in the set of all document words $R_1, \ldots, R_N$ to form a retrieval index. In the best possible case, this retrieval consists of all correct keywords available as its top results.

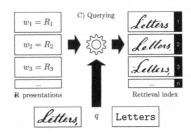

Fig. 2.6: Step (C): Querying of handwritten word representations.

In template-based KWS, a query word is directly matched against a set of document words in order to build the index. In learning-based KWS, the retrieval is based on a pre-trained statistical model. Both KWS approaches can be further distinguished with respect to their query input possibility. In the case of *query-by-example* (QbE), the user selects an exemplary word image of the keyword, while in the case of *query-by-string* (QbS), the user is able to retrieve words by arbitrary textual input. Finally, we also distinguish between segmentation-based and segmentation-free approaches. In the case of segmentation-free methods, the matching is conducted without any *a priori* segmentation, while in the case of segmentation-based approaches an explicit segmentation into line or word images is carried out before classification.

In the following section we start our review of template-based approaches for KWS and eventually, in Section 2.4.2, the review focuses on learning-based approaches.

2.4.1 *Template-based Approaches for KWS*

In the following two subsections, template-based matching algorithms for both statistical and structural representations are introduced.

2.4.1.1 *Statistical Representation*

In Table 2.6 various matching algorithms are summarised for statistical representations. These algorithms are then distinguished with respect to their segmentation level in Table 2.7. Finally, the different approaches are classified with respect to their query input possibility (i.e. QbE and QbS) in Table 2.8.

Early template-based KWS approaches are based on a pixel-by-pixel matching of a query image with all document images [45, 46, 100, 117, 122]. In particular, foreground and background pixels the size of normalised handwriting images are matched by a one-to-one correspondence. In the case of *XOR difference*, for instance, two images are combined by an XOR operation to retrieve the number of differing pixels [46, 100, 117, 122]. In the case of *Euclidean distance mapping* a similar approach is applied [45, 46, 100, 117, 122]. However, large pixel differences (e.g. *blobs*) are weighted more than small differences in this method. Another pixel-based approach is based on the *sum of squared differences* [46, 100, 117, 122], which measures the minimum translation cost between query and document image. However, holistic pixel-by-pixel matchings are easily affected by noise and other variations in handwriting, and thus, more elaborated approaches are based on the matching of more robust features.

In the case of sequences of feature vectors, we distinguish between *static* and *dynamic* matching approaches. In the former case, the i-th element of the sequence of feature vectors is assigned to the corresponding i-th element of the other sequence of feature vectors. In the latter case, the i-th element of a sequence of feature vectors can be assigned to an arbitrary element of the other sequence of feature vectors. Hence, static alignment is mostly employed for global and high-dimensional representations, while dynamic feature matching is employed for local or global but rather low-dimensional representations.

First dynamic approaches [45, 46, 100, 117] are based on an affine transformation of holistic contour points by means of the *Scott and Longuet-Higgins algorithm* [47]. Other contour-based approaches are based on the *shape distance* [159] that measures the dissimilarity of radial segmented contour points [46, 49, 100, 117] or a *multi-scale matching* approach for contours at different extraction levels [101]. In [143, 144], different approaches for the elastic and cohesive matching of gradient feature vectors are introduced.

A significant part of template-based matching algorithms is based on matching sequences of feature vectors rather than matching single holistic representations. That is, features are extracted by means of a sliding window that moves seamlessly over a word image to extract a set of different features at every window position. The result are sequences of feature vectors that can be optimally aligned by means of a dynamic programming approach such as, for example, *dynamic time warping* (DTW) (used, for instance, in [32, 57, 61, 110, 117]). Other dynamic programming approaches for this alignment are, for example, *continuous dynamic programming* [53, 115], *subsequence DTW* [115], or *flexible sequence matching* [115]. Other matching paradigms have also been applied in KWS, such as, for example, *RANSAC* [113], or *local proximity nearest neighbour search* [130].

In some cases feature vectors are not extracted by means of a sliding window approach, but instead, by a holistic or segmentation-based approach. In particular, low-dimensional features are often acquired for whole word images [125], around certain keypoints [104, 111, 146], or segments [124]. Often, these word images (or rather spatial segments) are represented by rather high-dimensional features such as HoG [54, 112], LBP [54, 55, 112], or BoVW [97, 131, 171]. Recently, these high-dimensional feature vectors have been embedded in low-dimensional feature spaces for KWS [116, 133]. In all of these cases, warping, as used in DTW for instance, is computationally too expensive due to the high dimensionality. Thus, static distance or similarity measures are often employed, such as, for example, *cosine* similarity [74, 97, 104, 125, 133], *Euclidean* distances [97, 104, 111, 116, 124, 131], *L1* and *L2* norms [54, 112, 133, 146], *Bray Curtis* [55, 133], or the χ^2 distance [171].

If we compare segmentation-free methods (i.e. page-based (e.g. [70, 112, 143, 146])) and segmentation-based methods (i.e. line- or word-based (e.g. [32, 45, 46, 52, 53, 57, 110, 116])), we observe that most of the template-based methods make use of full segmentations. One reason might be that

Table 2.6: Template-based matching algorithms for statistical representations.

Method	
Pixel-by-pixel	
- XOR difference	[46, 100, 117, 122]
- Euclidean distance mapping	[45, 46, 100, 117, 122]
- Sum of squared differences	[46, 100, 117, 122]
Feature (dynamic)	
- Scott and Longuet-Higgins	[45, 46, 46, 100, 117]
- Shape distance	[46, 49, 100, 117]
- Cohesive/elastic matching	[143, 144]
- Dynamic time warping	[32, 40, 42, 46, 50, 52, 56, 57, 61, 70, 99, 102, 110, 117, 122, 132]
- Continuous dynamic programming	[53, 115]
- Other	[101, 113, 115, 130]
Feature (static)	
- Cosine distance	[97, 104, 125, 133]
- Euclidean distance	[97, 104, 111, 116, 124, 131]
- L1 distance	[133]
- L2 distance	[54, 112, 133, 146]
- Bray Curtis	[55, 133]
- χ^2	[171]

template-based methods are more easily affected by subtle variations in the word images when compared to learning-based methods. Thus, segmentation and preprocessing is crucial for the accuracy of template-based methods. Moreover, the computational requirements are clearly increased when a query image needs to be matched against several prototype images rather than when two segmented word images need to be matched. In the case of segmentation-free approaches, we see that most approaches are based on detecting or segmenting a number of prototype word images or patches (i.e. candidates for document words). These candidate images are then used to match the query word graphs.

The vast majority of template-based approaches are based on a QbE approach (e.g. [32, 45, 46, 50, 53, 61, 98, 110, 144]). That is, a query word image is matched against a set of document images (or candidate images) to create a retrieval index. Vice versa, only few template-based approaches for QbS exist. In [144], for instance, a word image is synthetically generated based on a given textual query input.

Table 2.7: Classification of template-based matching algorithms for statistical representations by means of their segmentation levels.

Method	
Page-based	[70, 112, 113, 143, 144, 146, 171]
Line-based	[53, 99, 101]
Word-based	[32, 40, 42, 45, 46, 50, 52, 54–57, 61, 64, 76, 97, 98, 100, 102, 104, 109–111, 115–117, 122, 124, 125, 130–133]

Table 2.8: Classification of template-based matching algorithms for statistical representations by means of their query input.

Method	
QbE	[32, 40, 42, 45, 46, 49, 50, 52–57, 61, 70, 74, 97–102, 102, 104, 109–113, 115–117, 122, 124, 125, 130–133, 143, 144, 146, 171]
QbS	[144]

2.4.1.2 *Structural Representation*

In Table 2.9, various matching algorithms are summarised for structural representations. These algorithms are then distinguished with respect to their segmentation level in Table 2.10, and with respect to their query input possibility (i.e. QbE and QbS) in Table 2.11.

Most of the structural KWS approaches make use of a graph-based representation, and thus, graph matching is necessary for template-based KWS. Generally, graphs can be matched by means of *exact* or *inexact* graph-matching algorithms [173]. In contrast to exact graph matching, inexact graph matching is able to match graphs with variations in both their structure and labelling. In the case of graph-based KWS, graphs are used to represent handwriting and are thus affected by subtle variations. This makes exact graph matching infeasible, and thus, only inexact graph matching can be applied.

An early graph-based KWS approach [97] makes use of a *bag-of-paths* approach that is comparable to BoVW. In this approach all edges of a graph are compared with a set of prototype edges, and thus, a histogram with the number of prototype edge occurrences can be derived. The resulting histograms are then compared by means of the Euclidean distance.

More recent approaches try to directly match the structural representation rather than decomposing it. In [75, 129], for instance, the graph matching procedure is conducted in two steps. First, graph edit distances — a powerful and flexible inexact graph matching paradigm — between pairs of subgraphs are computed by means of a fast, but suboptimal algorithm [93]. Second, a minimum cost assignment of the individual subgraphs is found by means of DTW. The same suboptimal graph edit distance is also employed in [76] and [77]. However, in both cases the matching is conducted in a global manner, i.e. matching complete word graphs (rather than matching subgraphs).

In the case of graph-like representations as proposed in [108, 114], the matching is based on a *minimal energy function* that consists of two terms. The first term describes the deformation of the edges (named *springs* in this case), and the other describes the matching of the nodes (termed *inkballs* in this case). This particular matching is somehow comparable to graph edit distance as both approaches try to find an optimal assignment (i.e. a matching with minimal cost or energy) between substructures of the graphs.

Table 2.9: Template-based matching algorithms for structural representations.

Method	
Euclidean distance	[97]
Suboptimal graph edit distance	[75–77, 129]
Minimal energy function	[108, 114]

The vast majority of the structural approaches for template-based KWS are based on matching segmented word images [75, 76, 97, 108, 114, 129]. Only one approach is actually in part segmentation-free [77]. In this paper, a graph-based line segmentation algorithm [174] is first used to segment document pages into lines of text. Next, possible candidate graphs are found by means of a hashing index. Finally, graph edit distances between query and candidate graphs are computed.

All but one of the reviewed structural approaches are based on a QbE approach [75–77, 97, 108, 114, 129]. An initial QbS approach is proposed in [114], where graph-like inkball models are synthetically generated based on textual inputs. Following this, the actual matching is then based on matching the generated query model with all document models.

Table 2.10: Classification of template-based matching algorithms for structural representations by means of their segmentation levels.

Method	
Page-based	-
Line-based	[77]
Word-based	[75, 76, 97, 108, 114, 129]

Table 2.11: Classification of template-based matching algorithms for structural representations by means of their query input.

Method	
QbE	[75–77, 97, 108, 114, 129]
QbS	[114]

2.4.2   Learning-based Approaches for KWS

In the following two subsections, learning-based matching algorithms for both statistical and structural representations are introduced.

2.4.2.1   Statistical Representation

In Table 2.12, various matching algorithms are summarised for statistical representations. These algorithms are then distinguished with respect to their segmentation level in Table 2.13. Finally, the different approaches are classified with respect to their query input possibility (i.e. QbE and QbS) in Table 2.14.

In contrast to template-based approaches, learning-based KWS is based on statistical models that have to be trained *a priori* for the actual spotting task. Most approaches to learning-based KWS are based on *hidden Markov model* (HMM) (e.g. [11, 40, 52, 58, 63, 105]). HMMs are based on Markov chains with hidden states. A Markov model is a statistical model used to predict a future state based on observations (probabilities of certain features) in the current state. However, in the case of HMM the type of observations is not defined *a priori*, but based on a learning task.

Early approaches to HMM-based KWS are based on *generalised hidden Markov models* (gHMMs) that are trained on character images, i.e. images of Latin [11] or Arabic [59] characters. However, character-based segmentations are often error-prone and labour-intensive. Thus, more recent HMM approaches for KWS are based on feature vectors of word images [58], which are processed, for example, by means of *continuous-HMMs* (e.g. [32,61,106]) or *semi-continuous-HMMs* [61, 105, 140], i.e. HMMs with a shared set of GMMs. In [32] a lexicon-free HMM-approach is introduced that does not depend on an external language model. However, it has been shown that the use of a language model is beneficial for the same lexicon-free HMM-approach [107]. More recently, HMMs have been used in combination with FV [60], BoVW [62, 175], WG [123], and deep neural networks [56, 63].

Other learning-based approaches are, for example, based on *neural networks* (NNs) [126] and, in particular, a *recurrent neural network* (RNN) [41, 57, 96, 102, 103]. In the case of an RNN, the NN is, for example, based on hidden layers of *bidirectional long short-term memory* (BLSTM) nodes [176]. BLSTM nodes are used to address the *vanishing gradient problem*, i.e. the problem that large changes in the value of input layers do not have a large influence on the output layers. To recover sequences of words, RNN make use of the *CTC token passing* algorithm based on the letter probabilities of the BLSTM.

Support vector machines (SVMs) have also been employed for learning-based KWS [69–71, 118, 127, 128, 145]. SVMs permit linear and non-linear classifications by maximising the margin between the hyperplane of the classes to be separated. Further learning-based approaches make use of *latent semantic analysis* (LSA) [177] to learn the underlying semantic relationships between textual and/or visual feature spaces [72–74]. Thus, the retrieval of both visual (i.e. QbE) and textual (i.e. QbS) queries can be conducted in the same feature space by arbitrary distance metrics.

Last but not least, we have observed a clear shift towards *convolutional neural network* (CNN) for KWS in recent years (e.g [65, 66, 68, 170, 178]). CNNs are based on an input, an output, and several convolutional and pooling layers. Convolutional layers are used to apply a certain kernel (e.g. Gaussian blur, difference of Gaussians, etc.) to convolve an input image to an output image (termed *feature map*) with particular characteristics. This output image is then used as input for another convolutional layer, and so on. Each convolutional layer might consist of a different kernel function and parameters that are both learned during the training phase. Pooling layers are then used to average certain regions to eliminate variance with regard to scale, translation, and rotation. In some cases [64, 134], off-the-shelf CNNs are used as classifiers for KWS. However, in most cases, CNNs are used to learn certain word string embeddings such as PHOC [65–68,135,170,178], DCToW [66–68], SPOC [67], LSDE [68], and n-grams [66]. Thus, the KWS retrieval is then actually based on measuring the distance between query embedding and all document word embedding by means of a specific distance metric.

Table 2.12: Learning-based matching algorithms for statistical representations.

Method	
Hidden Markov model	[11, 32, 40, 52, 56, 58–63, 105–107, 120, 121, 123, 140]
Neural network	[126]
Recurrent neural network	[41, 57, 96, 102, 103]
Support vector machine	[69–71, 118, 127, 128, 145]
Latent semantic analysis	[72–74]
Convolutional neural network	[64–68, 134, 135, 170, 178]

Similar to template-based matching algorithms, most learning-based approaches make use of a full segmentation or at least a line segmentation of document images (e.g. [11, 40, 52, 58, 61, 65, 96, 105, 128]). That is, few approaches are actually segmentation-free (e.g. [62, 72, 145, 178]). Some of these approaches extract features on small patches over the whole image and then match them against a query [62, 70, 72, 74, 120, 145]. More recently, candidate images have first been detected and then used for the actual retrieval [170, 178].

Table 2.13: Classification of learning-based matching algorithms for statistical representations by means of their segmentation levels.

Method	
Page-based	[62, 70, 72, 74, 120, 145, 170, 178]
Line-based	[11, 32, 40, 41, 57, 59, 63, 96, 102, 103, 106, 107, 118, 121, 123, 127]
Word-based	[52, 56, 58, 60, 61, 64–66, 68, 69, 71, 73, 105, 119, 126, 128, 134, 135]

Learning-based approaches, in contrast to template-based approaches, inherently support the input of both QbE and QbS. That is, due to the *a priori* learning step, not only certain characteristics but also the mapping between certain features and characters (i.e. ASCII or unicode) can be trained. Thus, most learning-based approaches are also well-suited for textual query inputs (e.g. [11, 40, 59, 65, 69, 73, 96]).

Table 2.14: Classification of learning-based matching algorithms for statistical representations by means of their query input.

Method	
QbE	[52, 58, 60–62, 64–67, 69–72, 74, 105, 106, 118, 120, 126, 128, 134, 135, 145, 170, 178]
QbS	[11, 32, 40, 41, 56, 57, 59, 63, 65–69, 73, 96, 102, 103, 106, 107, 121, 123, 127, 128, 170]

2.4.2.2 *Structural Representation*

In Table 2.15, a first matching algorithm for structural representations is shown. This algorithm is then distinguished with respect to its segmentation level in Table 2.16. Finally, the same approach is classified with respect to its query input possibility (i.e. QbE and QbS) in Table 2.17.

Very recently, a learning-based approach for structural representations was presented in [179]. That is, graphs are enriched by means of two *message passing neural networks* (MPNNs) employed in a Siamese neural network architecture. In particular, MPNNs are used to learn the context of a node by sending messages through adjacent edges at different stages in time. As a result, node labels can be used to represent the learned context information. Following this, graph edit distances are computed between enhanced graphs by a fast, but suboptimal algorithm [79].

Table 2.15: Learning-based matching algorithms for structural representations.

Method	
Message passing neural network	[179]

Like all template-based matching algorithms for structural representations, the first learning-based approach makes use of a full segmentation of document images [179].

Table 2.16: Classification of learning-based matching algorithms for structural representations by means of their segmentation levels.

Method	
Page-based	-
Line-based	-
Word-based	[179]

In contrast to statistical learning-based approaches, querying has so far only been supported by means of QbS [179].

Table 2.17: Classification of learning-based matching algorithms for structural representations by means of their query input.

Method	
QbE	[179]
QbS	-

2.5 Summary

In recent decades, many handwritten historical documents have been digitised by different public or private organisations. However, many of these documents are affected by wide variations, and thus, an automatic full transcription is often not feasible. *Keyword spotting* (KWS) has been proposed to bridge the gap between the availability and accessibility of handwritten historical documents. KWS allows us to browse and search documents by arbitrary retrievals of given words. This chapter presents a broad survey of KWS research conducted in recent decades. The review gives an overview of past and upcoming trends in and around the topic of KWS.

Broadly speaking, KWS is based on three process steps. First, document images are preprocessed to reduce handwriting variations and conservation-related issues. In the same step, document pages are often segmented into smaller parts, e.g. lines of text or word images.

Based on preprocessed images, certain characteristics are extracted in the second step and stored in a specific representation formalism. In many cases, this formalism is based on a statistical representation by means of sequences of feature vectors. To this end, a sliding window approach (or similar) is used to extract features from the handwriting image. Broadly speaking, we distinguish between two types of statistical feature descriptors, handwriting feature descriptors that represent certain characteristics of the handwriting, and more generic texture and shape descriptors.

Handwriting can also be represented by means of structural formalisms. In this case, the inherent topological characteristics of handwriting images is often represented by means of a graph. However, even though structural approaches offer some representational advantages, until now, few graph-based KWS approaches have been observed. One reason for this could be the high computational complexity of graph-based pattern recognition when compared to statistical approaches.

In the third process step of KWS, a given input query needs to be matched or classified in order to form a retrieval index. In principle the keyword retrieval is either based on template-based or learning-based algorithms. In the former case, a query image is directly matched against a set of document words. In the latter case, a particular statistical model is trained *a priori*. This statistical model can then be used to recognise specific characteristics during the actual keyword spotting task. In general, template-based approaches are known for their high flexibility and generalisability as the matching is independent of the underlying handwriting and language. Learning-based approaches are dependent on *a priori* learning based on a relatively large set of labelled training data. As a result, learning-based approaches are generally better suited to dealing with certain variations in the documents and lead, in general, to higher accuracy rates when compared to template-based approaches.

In recent years we have observed certain trends in the field of KWS. First, there is a tendency towards segmentation-free approaches, that is, approaches where no explicit segmentation of document pages into line or word images is necessary prior to the KWS. Moreover, we have also observed a clear shift in the learning-based domain towards *convolutional neural network* (CNN) approaches in conjunction with word string embeddings. CNNs allow both very high recognition accuracy and a seamless integration of visual and textual querying. However, one needs to keep in mind that the accuracy of CNN-based approaches often crucially depends on the size of the labelled training data. However, labelling of historical handwriting documents is often a labour- and time-intensive task, and limited by the size of the document. Finally, there is a certain tendency to use structural rather than statistical representations for KWS. However, only few approaches for structural and, in particular, graph-based KWS have been proposed so far. This is particularly interesting as graph-based approaches have clear advantages in their representational power when compared to vectorial representations. Moreover, several fast suboptimal graph matching algorithms have been proposed in recent years. These algorithms allow us to employ graph-based approaches also in domains with rather large graphs and where a large amount of matching is required. This book therefore explores the applicability and advantages of graph-based approaches in the field of KWS.

Chapter 3

Datasets

In recent decades a wide set of handwritten historical documents have been made publicly available by digital means. These documents cover a wide range of different languages, alphabets, and time ranges, and thus, build an important pillar of our cultural heritage. To make these documents available for searching and browsing, different *keyword spotting* (KWS) approaches have been proposed as discussed in the previous chapter.

In the current chapter, we focus on four different handwritten ancient manuscripts, namely *George Washington* (GW)[1], *Parzival* (PAR)[2], *Alvermann Konzilsprotokolle* (AK)[3], and *Botany* (BOT)[3]. These documents range from the 13th century to the late 18th century, and consist of both English and German documents. The four documents are then used in the remainder of this book for experimental evaluations.

This chapter is organised as follows: We first introduce and describe the four different manuscripts in detail in Section 3.1, while the employed image preprocessing steps as well as the document segmentation are thoroughly explained in Section 3.2.

3.1 Historical Manuscripts

In Fig. 3.1, small excerpts from all four manuscripts are shown. Certain signs of degradation can be observed in all four manuscripts, while variation in the writing styles is rather low. Note that *George Washington* (GW) and

---

[1] George Washington Papers at the Library of Congress, 1741-1799: Series 2, Letterbook 1, pp. 270-279 & 300-309, http://memory.loc.gov/ammem/gwhtml/gwseries2.html

[2] Parzival at IAM historical document database, http://www.fki.inf.unibe.ch/databases/iam-historical-document-database/parzival-database

[3] Alvermann Konzilsprotokolle and Botany at ICFHR2016 benchmark database, http://www.prhlt.upv.es/contests/icfhr2016-kws/data.html

Parzival (PAR) are two well-known manuscripts in the field of KWS [32, 49, 58], while *Alvermann Konzilsprotokolle* (AK) and *Botany* (BOT) are based on a recent KWS benchmark competition [80].

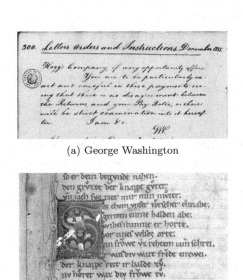

(a) George Washington

(b) Parzival

(c) Alvermann Konzilsprotokolle

(d) Botany

Fig. 3.1: Exemplary excerpts from the four historical manuscripts.

In Table 3.1, a summary of the characteristics of the four historical documents is provided. The oldest manuscript, PAR, dates back to the 13th century, while the newest document, BOT, was created in the 19th century. Two manuscripts are written in English and two written in German. Moreover, the variations with respect to style and size is rather low in all cases. That is, all manuscripts have been created by only a few writers. Note that in the case of AK and BOT the exact number of writers is not publicly available.

Table 3.1: Characteristics of the four historical documents.

Dataset	Century	Language	Handwriting variations	Writers
GW	18th	English	Low	2
PAR	13th	German	Low	3
AK	18th	German	Low	Unknown (few)
BOT	19th	English	Medium	Unknown (few)

The physical and digital formats of the four historical documents are shown in Table 3.2. Note that the exact writing tool is not known for any of the four documents. However, based on the time frame of the four manuscripts, we can safely assume that older documents such as PAR, GW and AK were written using a quill, while BOT was either written using a quill or a fountain pen. Moreover, three out of the four documents were written on paper, while only PAR was written on parchment. Note that the physical sizes of GW, AK and BOT are not publicly available.

Table 3.2: Page formats of the four historical documents.

Dataset	Physical format	Digital format
GW	Unknown size, ink on paper	2034 x 3286 px, JPEG, 300 dpi, greyscale
PAR	21.5 x 31.5 cm, ink on parchment	2000 x 3008 px, JPEG, 200 dpi, colour
AK	Unknown size, ink on paper	3200 x 5300 px, JPEG, 400 dpi, colour
BOT	Unknown size, ink on paper	3200 x 5300 px, JPEG, 400 dpi, colour

Finally, for all four historical documents we use a subset of the original document, as shown in Table 3.3[4]. The number of available words is equally distributed among the four manuscripts, except PAR. Broadly speaking,

---

[4]In the case of AK and BOT, we make use of the public available sets Train I and Test.

the complexity of a KWS evaluation increases with the number of available words. Note that the slightly uneven distribution is normalised by averaging the KWS performance over the employed queries.

Table 3.3: Number of available pages and words in the four historical documents.

Dataset	Available pages	Available words
GW	20	4,894
PAR	45	23,478
AK	30	5,383
BOT	30	4,914

In the following four subsections these documents are introduced in more detail.

3.1.1 George Washington (GW)

This document consists of twenty pages stemming from handwritten letters written by George Washington and his associates during the American Revolutionary War in 1755. The letters are written in English and contain only minor signs of degradation. The variation in the writing style is rather low, even though different writers were involved in their creation. However, there are noticeable intraword variations with respect to scaling (see Fig. 3.2).

3.1.2 Parzival (PAR)

This document consists of 45 pages stemming from handwritten letters written by the German poet Wolfgang Von Eschenbach in the 13th century. The manuscript is written in Middle High German on parchment with markable signs of degradation, as shown in Fig. 3.3. The variations in the writing are low, even though three different writers were involved (see Fig. 3.4).

3.1.3 Alvermann Konzilsprotokolle (AK)

This document consists of a subset of 18,000 pages stemming from handwritten minutes of formal meetings held by the central administration of the University of Greifswald in the period from 1794 to 1797. The notes

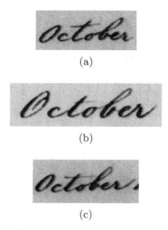

(a)

(b)

(c)

Fig. 3.2: Exemplary intraword variations of the word "October" in the George Washington manuscript. Note especially the different scalings of (a) and (c) when compared with (b).

are written in German and contain only minor signs of degradation. The variations in the writing styles are also rather low, as shown in Fig. 3.5.

3.1.4 Botany (BOT)

This document consists of a subset of of more than 100 different botanical records made by the government in British India in the period from 1800 to 1850. The records are written in English and contain certain signs of degradation and especially fading, smearing, and holes, as shown in Fig. 3.6. Note especially the vertical ruling lines in the background of Fig. 3.6a, which can be regarded as an additional challenge for KWS. The variations in the writing style are noticeable especially with respect to scaling and intraword variations, as shown in Fig. 3.7.

3.2 Preprocessing the Datasets

All four manuscripts are affected by variations, and thus, an image preprocessing is necessary. The processing steps applied in this book are discussed in this section.

Image preprocessing basically aims to reduce undesirable variations in the handwriting which are due to different writers (i.e. interpersonal variations, see, for example, Figs. 3.2, 3.4, 3.5, and 3.7) or the scanned document

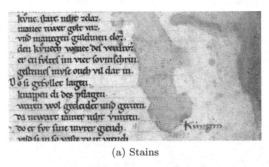

(a) Stains

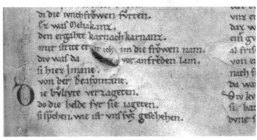

(b) Hole

Fig. 3.3: Exemplary signs of degradation in the Parzival manuscript.

itself (e.g. pixel noise, skewed scanning, or general degradation of the document, see, for example, Fig. 3.3, and 3.6). In our particular case, the preprocessing focuses on the latter case, as variations in the writing style are minimised by graph normalisations in a subsequent step (see Section 6.2.2 for details). Hence, the ultimate goal of the preprocessing is to make document images feasible for further graph extraction (see Chapter 4). That is, for the graph extraction we require single binarised and/or skeletonised word images. To this end, filtered and binarised document images need to be segmented into single skew corrected word images. Moreover, template-based KWS approaches, such as the proposed graph-based KWS framework, are negatively affected by noise. In order to achieve state-of-the-art results, it is crucial that variations (i.e. pixel noise, skew, etc) between word images of the same word class are minimised.

In Fig. 3.8, the image preprocessing process as well as small examples of its effects on the GW dataset are illustrated. Note that the whole image preprocessing process follows well-known techniques in document analysis as proposed by [51] and [141], which have proven successful for handwritten historical documents.

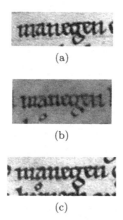

Fig. 3.4: Exemplary intraword variations of the word "manegen" in the Parzival manuscript. Note that only few variations can be observed in all three examples.

Fig. 3.5: Exemplary low intraword variations (style and size) of the word "worden" in the Alvermann Konzilsprotokolle manuscript. Note especially the small variations in the letter 'd' in (a) and (b) when compared with (c).

The complete process consists of six steps (A)-(F):

- (A) The first image preprocessing step locally enhances edges by means of *difference of Gaussians* (DoG) in order to address the issue of noisy background [141].
- (B) Next, filtered document images are binarised by a global threshold to clearly distinguish between the foreground (i.e. ink)

(a) Fading

(b) Smearing and hole

Fig. 3.6: Exemplary signs of degradation in the Botany manuscript.

and background (e.g. paper, parchment, leaves, etc.) of the hand-writing document.

- (C) On the basis of preprocessed document images, single word images are automatically segmented based on their projection profiles (GW and PAR) or by means of forced alignment (AK and BOT). Next, the segmentation result is manually inspected and, if necessary, manually corrected[5].
- (D) The skew, i.e. the inclination of the document, is estimated on the lower baseline of a line of text and then corrected on single word images [51].
- (E) In certain cases, the forced alignment segmentation is imperfect [80], and thus, additional filtering is necessary to remove undesired word parts. In particular, small connected components

---

[5]In the GW and PAR manuscripts, this KWS approach neglects any segmentation errors and can therefore be seen as an upper bound solution. In the AK and BOT manuscripts, however, the segmented word images are taken from the ICFHR2016 benchmark database, and thus, no manual segmentation correction is feasible. To mitigate the segmentation errors in these two documents, we employ additional morphological filtering (see step (E)).

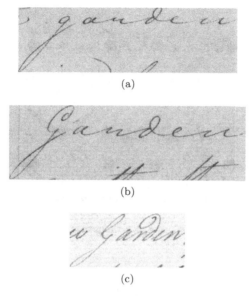

(a)

(b)

(c)

Fig. 3.7: Exemplary high intraword variations (in both style and scaling) of the word "Garden" in the Botany manuscript. Note especially the different style of the letter 'G'.

are detected and removed by means of different morphological operators.

- (F) Finally, binarised and normalised word images are skeletonised by a 3×3 thinning operator [180].

Note that we denote segmented word images that are binarised by B. If the image is additionally skeletonised we use the term S from now on.

Table 3.4 provides an overview of the employed image preprocessing steps for each manuscript. Note that not all image preprocessing steps have been conducted for all four different manuscripts. That is, in the GW and PAR manuscripts we conducted the full image preprocessing on our own, while in the AK and BOT manuscripts the image preprocessing is based on segmented word images of the ICFHR2016 benchmark database [80].

As a result, the word segmentation (step (C) in Table 3.4) has been conducted by two different approaches. Moreover, the skew is corrected for GW and PAR only (step (D) in Table 3.4), as no line images are provided for AK and BOT. In particular, it is not possible to accurately estimate skew on single word images, as cursive handwriting is characterised by large

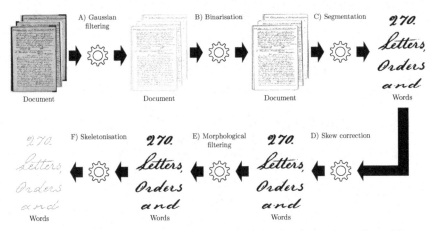

Fig. 3.8: Image preprocessing steps and their influence on document and word images in the George Washington document.

variations in its baseline (i.e. the curvature of single letters such as 'G' or 'g'). Finally, in the case of AK and BOT, no manual error correction of the segmentation results could have been conducted. Hence, additional morphological filtering is necessary to reduce the influence of imperfect segmentations (step (E) in Table 3.4).

In the next six paragraphs each image preprocessing step is described thoroughly. For each document and image preprocessing step, optimal parameter settings have been chosen visually. That is, the influence of certain image preprocessing steps has not been evaluated with respect to the KWS accuracy of the whole pipeline. This is due to the large number of potential combinatorial parameter settings of the complete framework. In terms of KWS accuracy, the current book focuses on the optimisation of parameters for graph representation and graph matching, while image preprocessing parameters are defined based on a visual inspection. This is certainly not an optimal solution, but rather, a feasible solution with respect to the large numbers of experiments that have been conducted (see Chapter 7).

3.2.1 Gaussian Filtering

To minimise the influence of noisy background (caused, for example, by stains, ink bleed-through, and the degradation of the document), we employ *difference of Gaussians* (DoG) filtering, as shown in Fig. 3.9. Given

Table 3.4: Conducted image preprocessing steps per manuscript.

Image preprocessing steps	GW	PAR	AK	BOT
A) Gaussian filtering	x	x	x	x
B) Binarisation	x	x	x	x
C) Segmentation (projection profiles)	x	x		
C) Segmentation (forced alignment)			x	x
D) Skew correction	x	x		
E) Morphological filtering			x	x
F) Skeletonisation	x	x	x	x

a particular input image, two filters with different degrees of blur are employed first. In particular, the difference between the more and less blurred image is derived to remove noise, or more precisely, to enhance edges. Formally,

$$I_{DoG} = I_{\sigma_2} - I_{\sigma_1} \quad ,$$

where I_{DoG} is the filtered output image and I_σ are blurred input images with different radii σ of the employed Gaussian blur filters. That is, the input image is convolved with two different Gaussian filters (i.e. small matrices) with different radii σ (and variance). For DoG, σ_2 is generally higher — and therefore the degree of blur is also higher — than σ_1.

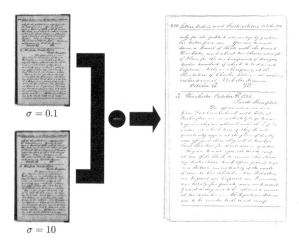

$\sigma = 0.1$

$\sigma = 10$

Fig. 3.9: An example of difference of Gaussians filtering with two different blur radii σ.

For the four different manuscripts, we employ three different radii for $\sigma_1 = \{0.1, 0.5, 1.0\}$ in combination with two radii for $\sigma_2 = \{10, 40\}$. Next, we manually inspect the filtering results and choose the optimal parameter settings, as shown in Table 3.5.

Table 3.5: Optimal parameter settings for difference of Gaussians filtering per manuscript.

Manuscript	σ_1	σ_2
GW	0.1	10
PAR	1.0	10
AK	0.5	40
BOT	1.0	40

3.2.2 *Binarisation*

To distinguish between foreground and background, the filtered document images are binarised by means of global thresholding as illustrated in Fig. 3.10. That is, every pixel i of a greyscale image I is either assigned to the foreground or background depending whether the intensity of the pixel $\eta(i)$ is below or above a certain (global) threshold $\mu \in [0, 255]$. Hence, initially every pixel i has an intensity between 0 and 255 where $\eta(i) = 0$ refers to black and $\eta(i) = 255$ refers to white. For our binarisation, every pixel i in a greyscale image I is then transformed to

$$\hat{\eta}(i) = \begin{cases} 0 \text{ (foreground/ink)}, & \text{if } \eta(i) < \mu \\ 1 \text{ (background)}, & \text{otherwise} \end{cases}$$

where $\hat{\eta}(i)$ is the *posteriori* binary value of pixel i, and μ is the threshold value. Generally speaking, if μ is large, more pixels are assigned to the foreground, and likewise, if μ is small, more pixels are assigned to the background.

The optimal value of μ varies for every manuscript and can be estimated manually or automatically (i.e. by automatically thresholding the greyscale intensity histogram as illustrated in Fig. 3.8). In the current case, the results obtained with three different parameter settings $\mu = \{210, 220, 230\}$ are manually inspected. It turns out that a global threshold of $\mu = 220$ leads to good results on all four manuscripts.

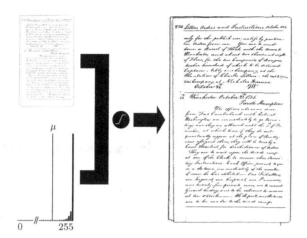

Fig. 3.10: Binarisation of document images by means of a global threshold μ.

3.2.3 *Segmentation*

Next, single word images are extracted (semi-)automatically based on the preprocessed and binarised document images. The extraction of word images from document images is a two-step procedure. First, lines of text are extracted from every document image, and second, single word images are extracted from the isolated text lines. Note that we only conduct the segmentation for GW and PAR, while AK and BOT are segmented in the context of the ICFHR2016 benchmark database [80].

For GW and PAR, lines of text are first automatically segmented and, if necessary, manually corrected [181]. Next, the vertical projection profile of a line of text is used to automatically segment word images as illustrated in Fig. 3.11. To this end, a histogram of the vertical projection profile $H = \{h_1, \ldots, h_N\}$ is built. The bins $h_i \in H$ count the number of foreground pixels in column i of a text line image. Next, this line of text is automatically segmented in every white space position (i.e. in positions where $h_i = 0$) (see the dotted lines in Fig. 3.11). However, this segmentation procedure can lead to over- and under-segmentation (see, for instance, the words *270.* and *Instructions.* in Fig. 3.11). Thus, manual error correction is applied in this book.

For AK and BOT, lines of text are first manually segmented and then further segmented into word images by means of forced alignment [80]. That is, candidate word images are extracted and aligned by means of the word length of a full transcription. However, this is a rather error-prone approach, and thus we apply further filtering (see Subsection 3.2.5). In particular, the length of handwritten words often does not fully correspond with the length of a textual transcription. Note that manual error correction is often not feasible, especially in cases of over-segmentation, where parts of the word image are missing.

Fig. 3.11: Segmentation of lines of text into word images by means of projection profiles.

3.2.4 *Skew Correction*

Once single word images are extracted, the skew is estimated on lines of text and then corrected on single word images as illustrated in Fig. 3.12. Note that this correction has only been employed for GW and PAR, as no text line images are publicly available for AK and BOT [80]. In Fig. 3.12a, a certain skew is observable when compared to the lower baseline (visualised with a dotted black line). To correct this variation, we estimate the skew angle by means of a linear regression on the lower baseline of a line of text [51]. In particular, we first obtain the lower baseline of a line of text $P = \{p_1, \ldots, p_N\}$ where p_i is the (x, y)-position of the first foreground pixel from the bottom of column i of the line of text, see Fig. 3.12b. However, the obtained lower baseline, P, is affected by outliers, and thus, error correction is needed prior to the actual skew correction.

To this end, we first estimate the lower baseline by means of linear regression. Formally,

$$y = ax + b \quad ,$$

$$a = \frac{\sum_{i=1}^{N} x_i y_i - N \mu_x \mu_y}{\sum_{i=1}^{N} x_i^2 - N \mu_x^2} \quad ,$$

$$b = \mu_y - a \mu_x \quad ,$$

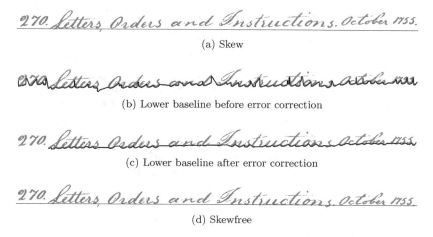

(a) Skew

(b) Lower baseline before error correction

(c) Lower baseline after error correction

(d) Skewfree

Fig. 3.12: Skew estimation and correction of lines of text by means of linear regression.

where (x_i, y_i) is the position of the lowest foreground pixel in column i, and μ_x and μ_y are the means of all x- and y-positions, respectively. Next, the sum of squared errors e of the linear regression is computed. Formally,

$$e = \sum_{i=1}^{N}(ax_i + b - y_i)^2 \quad .$$

If e is above a certain threshold t_e, the baseline point p_i with the largest error is removed from P. This procedure is repeated until e is below a user-defined threshold t_e[6]. Once the sum of squared errors is sufficiently small (see Fig. 3.12c), the skew angle α is estimated by $\alpha = \text{atan}(a)$ and then corrected on word images by means of a rotation of $-\alpha$ (see Fig. 3.12d).

3.2.5 *Morphological Filtering*

In order to reduce segmentation errors, we employ additional filtering based on morphological operators as illustrated in Fig. 3.13. This process is similar to DoG. Note that this procedure is employed on AK and BOT only as manual and uncorrected line segmentation in combination with the word segmentation based on forced alignment leads to markable segmentation errors (see for instance Fig 3.13a).

---

[6]We tested different error thresholds $t_e = \{5, \ldots, 75\}$ and chose the thresholds manually for GW and PAR. These are $t_e = 35$ and $t_e = 45$, respectively.

(a) Unfiltered word images

(b) Blurred and filled word images

(c) Largest connected component

(d) AND filtering

(e) Small connected component filtering

Fig. 3.13: Morphological filtering to reduce segmentation errors on Alvermann Konzilsprotokolle (left) and Botany (right).

First, we remove very small connected components with a foreground pixel count of less than 10. Next, a morphological blurring by means of *dilatation*[7] and morphological filling by means of *closing*[8] are employed as illustrated in Fig. 3.13b [182]. In addition, we insert a synthetic baseline on the BOT manuscript, as the handwriting is often not coherent. We then extract the connected component with the highest foreground pixel count from the resulting word images, as illustrated in Fig. 3.13c. This word image can then be used as an AND filter with the initial word image (see Fig. 3.13a). As a result, all foreground pixels that do not belong to the actual word are removed, as shown in Fig. 3.13d. Lastly, small connected components with a foreground pixel count below 50 are removed, as shown in Fig. 3.13e.

---

[7]Available by Matlab's `imdilate`. We define a 5×5 and 8×8 blurring matrix for AK and BOT, respectively.

[8]Available by Matlab's `imclose`. We define a a 4×4 filling matrix for both AK and BOT, respectively.

3.2.6 *Skeletonisation*

Finally, the preprocessed and binarised word images are skeletonised (also termed *thinned*), as illustrated in Fig. 3.14. That is, foreground pixels at the boundaries of a binary image are iteratively removed by means of thinning[9] until no further boundary pixels can be removed. The remaining pixels can be seen as the skeleton of the original image. Hence, this operation can be interpreted as a reduction of the handwriting to a one-pixel-wide representation of strokes.

(a) George Washington

(b) Parzival

(c) Alvermann Konzilsprotokolle

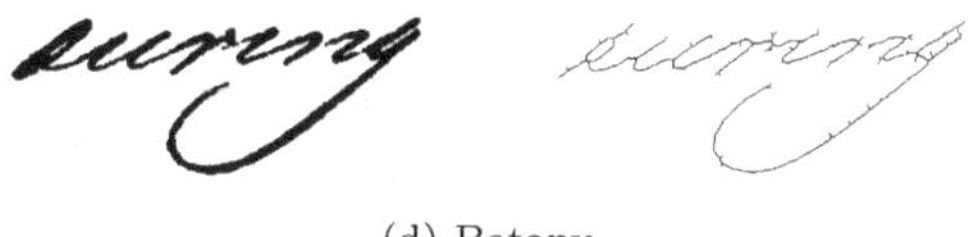

(d) Botany

Fig. 3.14: Preprocessed and binarised word images (left) as well as their skeleton images (right).

However, in certain cases where the contour of word images (i.e. the strokes) are not smooth, we can observe certain skeletonisation errors, as shown in Fig. 3.15. These effects lead to small "spikes" along the skeletonisation that might influence further graph representation. In particular, this occurs on Keypoint graphs (see Section 4.3.1), in which nodes are represented by means of characteristic points on the skeleton, and edges are used to represent strokes between these characteristic points. However, we need

---

[9]Available by Matlab's bwmorph with a 3 × 3 thinning matrix.

to consider that only a small fragment of nodes is based on these spikes, and thus, for further evaluation we decide not to employ further image pre-processing to mitigate this effect[10]. In the remaining graph representations, these imperfect skeletonisations have no effect, as nodes are represented by means of the centre of mass of segments of binarised, but not skeletonised, images.

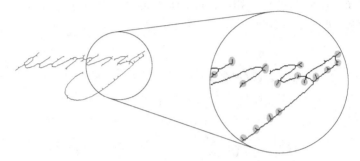

Fig. 3.15: Imperfect skeletonisation by means of small spikes (highlighted in grey).

3.3 Summary

To make handwritten historical documents accessible for browsing and searching, keyword spotting has been proposed as an alternative to an automatic full transcription. In the current book, four different ancient manuscripts are researched, namely *George Washington* (GW), *Parzival* (PAR), *Alvermann Konzilsprotokolle* (AK), and *Botany* (BOT). These Latin documents cover a time range from the 13th to the 18th century. In GW, stories of George Washington and his associates during the American Revolutionary War are journalised, while PAR is based on handwritten letters of the German poet Wolfgang Von Eschenbach. Handwritten minutes of formal meetings of the University of Greifswald are recorded in the AK manuscript, while different botanical records are described by the British India government in BOT. In all four manuscripts we observe certain signs of degradation, while the variation in handwriting is rather low.

Image preprocessing is necessary to reduce the influence of these effects. Moreover, the keyword spotting framework researched in this book makes

---

[10]One might, for example, split the connected components to connected subcomponents by removing junction points. Following this, small connected subcomponents (i.e. spikes) where the foreground pixel count is below a certain threshold could be removed.

use of segmented word images, and thus, the segmentation of the scanned document images is also required. In particular, the first image preprocessing step enhances edges locally by means of *difference of Gaussians* (DoG) in order to address the issue of noisy background [141]. That is, an image is first convolved by two different Gaussian matrices. Next, the difference between these two blurred images is used to enhance edges and filter noise. In the second image preprocessing step, the filtered document images are binarised by a global threshold to clearly distinguish between the foreground (i.e. ink) and background (e.g. paper, parchment, etc.) in the handwritten documents. Next, single word images are automatically segmented, based on the preprocessed document images. That is, lines of text are automatically segmented into word images by means of white spaces in the vertical projections profiles or by means of the forced alignment of a full transcription. The segmentation result is inspected manually and, if necessary, corrected (this step is only employed for GW and PAR). In the next image preprocessing step, the skew, i.e. the inclination of the document, is estimated on the lower baseline of a line of text and then corrected on single word images [51]. Moreover, additional filtering is needed as the forced alignment segmentation of AK and BOT leads to undesired effects. In particular, small connected components that do not belong to the segmented word images are removed by means of different morphological operators. Finally, binarised and normalised word images are skeletonised by a 3×3 thinning operator.

Chapter 4

Graph Representations

4.1 Introduction and Basic Definitions

Pattern recognition applications are either based on statistical (i.e. vectorial) or structural data structures (i.e. strings, trees, or graphs). Hence, *keyword spotting* (KWS) approaches also make use of either a statistical or a structural representation of handwriting, as reviewed in Subsection 2.3.1 and 2.3.2, respectively.

Graphs, in contrast to feature vectors, are able to represent both entities and binary relationships that might exist between these entities. Moreover, graphs can adapt their size and complexity to the size and complexity of the actual pattern to be modelled [183]. Due to their representational power and flexibility, graphs have found widespread application in pattern recognition and related fields. Prominent examples of classes of patterns which can be formally represented in a more suitable and natural way by means of graphs rather than with feature vectors are chemical compounds [184], documents [185], proteins [186], and networks [187] (see [87, 188, 189] for a survey on applications of graphs in pattern recognition). However, only very few applications can be found [75–77, 119] where graphs have been used to represent handwriting images. This is rather surprising as the inherent power and flexibility of graphs is particularly well-suited to representing handwriting. That is, graphs are able to adapt their structure to the variability in the underlying handwriting. Moreover, graphs are able to represent binary relationships in the structure of the handwriting. Overall, graphs can be seen as a very natural way to represent handwriting.

For this reason, this chapter focuses on graph representations for handwriting. That is, different formalisms for the representation of handwriting by means of graphs are introduced. In general, all formalisms aim to extract the minimal amount of topological information needed to preserve the inherent characteristics of the handwritten word.

An initial approach makes use of certain characteristic points (so-called keypoints) that are represented by nodes, while edges represent strokes between these points. Another approach is based on a grid-wise segmentation of word images, where each segment is represented by a node. Finally, two representation formalisms are based on vertical and horizontal segmentations of word images by means of projection profiles.

The remainder of this chapter is organised as follows: The formal concept of a graph and related concepts are introduced in Section 4.2. Next, four different graph representations are proposed in Section 4.3. Finally, these formalisms are evaluated in four different ancient manuscripts in a classification experiment in Section 4.4.

4.2 Graph-Based Pattern Recognition

In the following, a formal introduction to graphs and related concepts is provided.

Definition 4.1. (*Graph*) A graph g is defined as a four-tuple $g = (V, E, \mu, \nu)$ where

- V is a finite set of nodes
- E is a finite set of edges
- $\mu : V \to L_V$ is a node labelling function, and
- $\nu : E \to L_E$ is an edge labelling function

Given this definition, graphs can be characterised by means of the direction of their edges as well as their labelling functions. In particular, graphs can be divided into *undirected* and *directed* graphs, where pairs of nodes are either connected by undirected or directed edges, respectively. Additionally, graphs are often distinguished into *unlabelled* and *labelled* graphs. In the latter case, both nodes and edges can be labelled with an arbitrary numerical (i.e. $L = \{1, 2, 3, \ldots\}$), vectorial (i.e. $L = \mathbb{R}^n$), or symbolic label (i.e. $L = \{\alpha, \beta, \gamma, \ldots\}$) from L_V or L_E, respectively. In the former case we assume empty label alphabets, i.e. $L_V = L_E = \varnothing$.

Different types of graphs are illustrated in Fig. 4.1. In particular, undirected graphs are shown in Fig. 4.1a and 4.1c, while directed graphs are shown in Fig. 4.1b and 4.1d. Moreover, unlabelled graphs are shown in Fig. 4.1a and 4.1b, while labelled graphs are shown in

Fig. 4.1c and 4.1d. In the latter case, L_V is defined by symbolic labels {light grey, dark grey, black} while L_E is first defined by ∅ and then replaced by symbolic labels i.e. $L_E = \{a, b, c, d, e, f\}$.

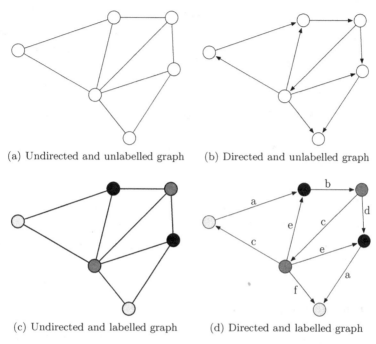

(a) Undirected and unlabelled graph   (b) Directed and unlabelled graph

(c) Undirected and labelled graph   (d) Directed and labelled graph

Fig. 4.1: Different types of graphs (w.r.t. direction of edges and labelling).

A subgraph shares most nodes and edges (including their labelling) of a certain parent graph, similar to the subset relationship in set theory [183]. That is, a subgraph is formally defined as follows:

Definition 4.2. (*Subgraph*) A subgraph $g_1 = (V, E, \mu, \nu)$ of $g_2 = (V, E, \mu, \nu)$ (denoted by $g_1 \subseteq g_2$) is defined by

(1) $V_1 \subseteq V_2$
(2) $E_1 \subseteq E_2$
(3) $\mu_1(u) = \mu_2(u)$ for all $u \in V_1$, and
(4) $\nu_1(e) = \nu_2(e)$ for all $e \in E_1$

If the second condition is replaced by the following more stringent condition,

$$(2') \quad E_1 = E_2 \cap V_1 \times V_1 \quad,$$

g_1 becomes an *induced* subgraph of g_2.

That is, g_1 is an induced subgraph of g_2 when some nodes of g_2 and their adjacent edges are removed (but no further edges are removed). Exemplary subgraphs are shown in Fig. 4.2, where an induced subgraph (see Fig. 4.2b) as well as a non-induced subgraph (see Fig. 4.2c) is shown.

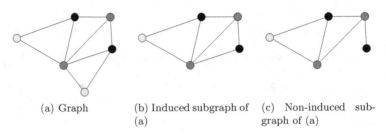

(a) Graph   (b) Induced subgraph of (a)   (c) Non-induced subgraph of (a)

Fig. 4.2: In Subfigure (b) and (c) we show different types of subgraphs of the graph shown in Subfigure (a).

4.3   From Handwriting to Graphs

On the basis of segmented, binarised, and possibly skeletonised word images (see Chapter 3), we now define four graph representation formalisms. These formalisms aim to represent certain characteristics of a word image by means of a graph. These handwriting graph representations will later be used to spot a query word (represented by a query graph) in a given document (represented by a set of document graphs). For all graph representations defined in the following four subsections, nodes are labelled with two-dimensional numerical labels, while edges remain unlabelled, i.e. $L_V = \mathbb{R}^2$ and $L_E = \varnothing$.

For every graph representation, we visualise the influence of different parameter values on graphs with three different sizes, namely small, medium, and large. The parameter values of each graph representation are chosen in a way that graphs with the target graph sizes, as indicated in Table 4.1, are extracted from the manuscripts.

This procedure allows us to visually compare the influence of different parameter settings with respect to both size and topological characteristics.

Table 4.1: Definition of graph sizes.

Size	Mean number of nodes
Small	25 to 50
Medium	50 to 75
Large	75 to 100

Further graph visualisations for all graph representation formalisms and manuscripts can be found in Chapter 9.

4.3.1 *Keypoint-Based Graphs*

The first graph representation is based on so-called keypoint graphs, which have been introduced for handwriting recognition in [119]. Graphs are extracted based on the detection of characteristic points, or keypoints, in binarised and skeletonised word images S. Keypoints can be seen as the minimum number of points needed to keep the inherent topological characteristic of a word image.

The proposed procedure extends and refines the algorithm introduced in [119]. It is formalised in Algorithm 1 (denoted by Keypoint from now on). First, end points and junction points are identified for each *connected component* (CC) of the skeleton image (see line 2 of Algorithm 1 and Fig. 4.3a). In particular, end points are detected by searching foreground pixels with only one neighbouring foreground pixel[1], while junction points are detected by searching foreground pixels with at least three or more neighbouring foreground pixels. For circular structures, such as for instance the letter 'O', the upper left point of the skeleton is selected as a junction point. Note that most keypoints are either based on end points or junction points with four neighbouring foreground pixels. However, especially in the case of a flourish at the beginning or end of a word (see Fig. 4.3), we can observe junction points with only three neighbouring foreground pixels. For this reason, it is crucial to also cover such junction points. Selected points are added to the graph as nodes and labelled with their (x, y)-coordinates (see line 3).

Note that the skeletons based on [180] may contain several neighbouring end points or junction points, from which only one specific point is selected

---

[1]In certain cases, end points can also have more than one neighbouring foreground pixel, i.e. in the case of "L"- or "T"-shaped foreground pixel structures.

by means of a local search procedure (this step is not explicitly formalised in Algorithm 1). To this end, we extract only junction points within a certain neighbourhood that maximise the number of neighbouring foreground pixels, while end points that minimise the number of neighbouring foreground pixels are selected. Basically, a junction point is selected as a node if no other junction point exists in the 8-connected neighbourhood with higher connectivity (i.e. more neighbouring foreground pixels). In the case of end points, only end points that have no other end point in their 8-connected neighbourhood with lower connectivity are selected as nodes.

Next, the junction points are removed from the skeletonised image. This removal divides the skeleton into *connected subcomponents* (CC_{subs}) (see line 4 of Algorithm 1 and Fig. 4.3b). For each connected subcomponent, intermediate points $(x, y) \in CC_{sub}$ are converted to nodes and added to the graph in equidistant intervals of size D (see line 5 and 6 of Algorithm 1 and Fig. 4.3c). If an intermediate point is adjacent to a junction point, it is merged with the corresponding node of the junction point (this step is not explicitly formalised in Algorithm 1).

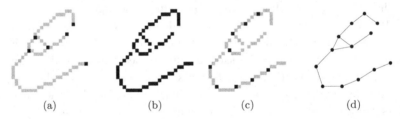

(a)   (b)   (c)   (d)

Fig. 4.3: Graph representation `Keypoint` for the letter "e". First, end points and junction points are detected in Subfigure (a). Next, junction points are removed to create connected subcomponents in Subfigure (b). Further intermediate points are then detected in Subfigure (c). Finally, a graph created by means of the detected keypoints is shown in Subfigure (d).

Finally, an undirected edge (u, v) between $u \in V$ and $v \in V$ is inserted into the graph for each pair of nodes that is directly connected by a chain of foreground pixels in the skeleton image S (see line 7 and 8 of Algorithm 1 and Fig. 4.3d). Note that this procedure fails if there are holes in the word images, as shown in Fig. 4.3d. In order to mitigate this problem, one could, for example, add an edge between near end points. However, in the current book we do not consider such an additional processing step.

Algorithm 1 Graph extraction based on keypoints

In: Skeleton image S, Distance threshold D
Out: Graph $g = (V, E)$ with nodes V and edges E
1: **function** KEYPOINT(S,D)
2:     **for** Each connected component $CC \in S$ **do**
3:         $V = V \cup \{(x, y) \in CC \mid (x, y) \text{ are end points or junction points}\}$
4:         Remove junction points from CC
5:         **for** Each connected subcomponent $CC_{\text{sub}} \in CC$ **do**
6:             $V = V \cup \{(x, y) \in CC_{\text{sub}} \mid (x, y) \text{ are points in equidistant intervals } D\}$
7:     **for** Each pair of nodes $(u, v) \in V \times V$ **do**
8:         $E = E \cup (u, v)$ if the corresponding points are connected in S
9:     **return** g

Fig. 4.4 shows examples of graph representations of four different handwritten words in three different sizes (small, medium, large), stemming from the four historical manuscripts. Note especially that the number of intermediate nodes increases with larger graphs, making the graphs' topological characteristics visually more similar to the original handwriting. In Chapter 9 in Tables 9.1, 9.5, 9.9, and 9.13, further graph representations can be found for every manuscript.

4.3.2   *Grid-Based Graphs*

The second graph representation is based on a grid-wise segmentation of binarised word images B (i.e. in contrast with Keypoint, no skeletonisation is applied).

The procedure is formalised in Algorithm 2 (denoted by Grid from now on). First, the dimensions of the segmentation grid are derived, basically defined by the number of columns, C, and rows, R (see lines 2 and 3 of Algorithm 2 and Fig. 4.5a and 4.5b). Formally, we compute

$$C = \frac{\text{Width of } B}{w} \text{ and } R = \frac{\text{Height of } B}{h} \quad,$$

where w and h denote a user-defined width and height of the resulting segments.

Next, a word image, B, is divided into $C \times R$ segments of equal dimension. For each segment s_{ij} ($i = 1, \ldots, C; j = 1, \ldots, R$) a node is inserted into the resulting graph and labelled by the (x, y)-coordinates of the centre

Original	Preprocessed	Small	Medium	Large

(a) George Washington

(b) Parzival

(c) Alvermann Konzilsprotokolle

(d) Botany

Fig. 4.4: Original and preprocessed word images, as well as corresponding Keypoint graph representations stemming from different manuscripts.

of mass (x_m, y_m) (see line 4). Formally, for each segment we compute

$$x_m = \frac{1}{n} \sum_{w=1}^{n} x_w \text{ and } y_m = \frac{1}{n} \sum_{w=1}^{n} y_w \quad ,$$

where n denotes the number of foreground pixels in segment s_{ij}, while x_w and y_w denote the x- and y-coordinates of these foreground pixels in s_{ij}. If a segment does not contain any foreground pixels, no centre of mass can be determined and thus no node is created for this segment.

Finally, undirected edges (u, v) are inserted into the graph in two subsequent steps. First, edges are inserted according to the four neighbouring segments on the top, left, right, and bottom of a node $u \in V$. If a neighbouring segment of u is also represented by a node $v \in V$, an undirected edge (u, v) between u and v is inserted into the graph (see Fig. 4.5c). Second, the resulting edges are reduced by means of a *minimal spanning tree* (MST) algorithm [190]. Hence, the graphs are transformed into trees (see Fig. 4.5d).

(a)   (b)   (c)   (d)

Fig. 4.5: Graph representation Grid for the word "Orders". First, the binarised word image is segmented in a vertical and horizontal direction in Subfigure (a) and (b), respectively. Next, a graph is created based on neighbouring segments in Subfigure (c). Finally, this graph is reduced to a tree by means of a minimal spanning tree algorithm in Subfigure (d).

Algorithm 2 Graph extraction based on a segmentation grid

In: Binary image B, Grid width w, Grid height h
Out: Graph $g = (V, E)$ with nodes V and edges E
1: **function** GRID(B,w,h)
2:     **for** $i \leftarrow 1$ to number of columns $C = \frac{\text{Width of } B}{w}$ **do**
3:         **for** $j \leftarrow 1$ to number of rows $R = \frac{\text{Height of } B}{h}$ **do**
4:             $V = V \cup \{(x_m, y_m) \mid (x_m, y_m)$ is the centre of mass of segment $s_{ij}\}$
5:     **for** Each pair of nodes $(u, v) \in V \times V$ **do**
6:         $E = E \cup (u, v)$ if associated segments are adjacent and connected by MST
7:     **return** g

Fig. 4.6 shows examples of graph representations of the four different handwritten words in three different sizes, stemming from the four historical manuscripts. Note in particular that certain visual characteristics of the handwriting are missing in the case of small graphs (see, for example, Fig. 4.6a). Moreover, it is observable that the edges are not coherent with the actual handwriting strokes. In particular, circular structures are often not closed due to the edge reduction by means of MST. In Chapter 9 in Tables 9.2, 9.6, 9.10, and 9.14, further graph representations can be found for every manuscript.

4.3.3 *Projection-Based Graphs*

The next graph representation is based on an adaptive rather than a fixed segmentation of binarised word images B. Similar to *line adjacency graphs* [191] and *run graphs* [192–194], nodes are represented by vertical and/or horizontal foreground pixel segments, while edges are used to represent the adjacency between neighbouring segments. That is, the individual segment sizes are adapted with respect to vertical and horizontal projection profiles.

The procedure is formalised in Algorithm 3 (denoted by Projection

Original	Preprocessed	Small	Medium	Large

(a) George Washington

(b) Parzival

(c) Alvermann Konzilsprotokolle

(d) Botany

Fig. 4.6: Original and preprocessed word images, as well as corresponding Grid graph representations stemming from different manuscripts.

from now on). First, a histogram of the vertical projection profile $P_v = \{p_1, \ldots, p_{max}\}$ is computed, where p_i represents the frequency of foreground pixels in column i of image B and max is the width of B (see line 2 of Algorithm 3). Next, we split B vertically by searching so-called white spaces, i.e. subsequences $\{p_i, \ldots, p_{i+k}\}$ with $p_i = \ldots = p_{i+k} = 0$. We split B in the middle of all white spaces, i.e. at position $p = \lfloor (p_i + p_{i+k})/2 \rfloor$. This procedures results in n segments $\{s_1, \ldots, s_n\}$ (see line 3 of Algorithm 3 and Fig. 4.7a). In the best case a segment encloses word parts that semantically belong together (e.g. characters)[2]. If the width of a segment $s_i \in \{s_1, \ldots, s_n\}$ is greater than D_v, s_i is further subdivided into subsegments with equidistant width D_v (see lines 4 and 5 of Algorithm 3 and Fig. 4.7b).

Based on the projection profile of rows (rather than columns) the same procedure as described above is then applied to each (vertical) segment $s_i \in \{s_1, \ldots, s_n\}$ (rather than to complete word images B) (see lines

---

[2]By means of this splitting procedure, we guarantee that nodes — inserted at the centre of mass of a segment — represent (sub)parts of whole characters only, i.e. no nodes are inserted for segments that represent different characters.

(a)　　　　　　　　　　(b)

Fig. 4.7: Vertical projection profile of the word "Orders". First, the binarised word image is segmented vertically in white space in the projection profile in Subfigure (a). Next, further segments are created at equidistant intervals in Subfigure (b).

6 to 10 of Algorithm 3 and Fig. 4.8). Thus, each vertical segment s_i is divided individually into horizontal segments $\{s_1, \ldots, s_m\}$. Note that a user-defined parameter, D_h, controls the number of additional segmentation points (similar to D_v).

(a)　　　　　　　　　　(b)

Fig. 4.8: Horizontal projection profile of segments of the word "Orders". First, the binarised word image is segmented horizontally in white space in the projection profile in Subfigure (a). Next, further segments are created at equidistant intervals in Subfigure (b).

Subsequently, for each final segment, s, a node is inserted into the resulting graph and labelled with the (x, y)-coordinates of the centre of mass (x_m, y_m) of this segment (see lines 11 and 12). If a segment consists of background pixels only, no centre of mass can be determined and thus no node is created for this segment.

Finally, an undirected edge (u, v) between $u \in V$ and $v \in V$ is inserted into the graph for each pair of nodes, if the corresponding pair of segments is directly connected by a chain of foreground pixels in the skeletonised word image S (see lines 13 and 14 and Fig. 4.9). In particular, an edge is either inserted if two segments are connected by a chain of foreground pixels in a vertical (see Fig. 4.9a), horizontal (see Fig. 4.9b), or diagonal (see Fig. 4.9c) direction.

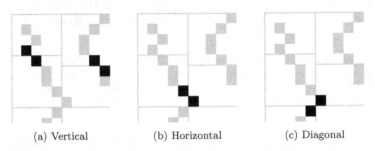

| (a) Vertical | (b) Horizontal | (c) Diagonal |

Fig. 4.9: Edge insertion based on neighbouring segments in a vertical (see (a)), horizontal (see (b)), or diagonal (see (c)) direction.

Algorithm 3 Graph extraction based on projection profiles

In: Binary image B, Skeleton image S, Vertical threshold D_v, Horizontal threshold D_h
Out: Graph $g = (V, E)$ with nodes V and edges E
1: **function** PROJECTION(B,S,D_v,D_h)
2:     Compute vertical projection profile P_v of B
3:     Split B vertically at middle of white spaces of P_v into $\{s_1, \ldots, s_n\}$
4:     **for** Each segment $s_i \in \{s_1, \ldots, s_n\}$ with width larger D_v **do**
5:         Split s_i vertically in equidistant intervals D_v
6:     **for** Each segment $s_i \in \{s_1, \ldots, s_n\}$ **do**
7:         Compute horizontal projection profile P_h of s_i
8:         Split s_i horizontally at middle of white spaces of P_h into $\{s_1, \ldots, s_m\}$
9:         **for** Each segment $s_j \in \{s_1, \ldots, s_m\}$ with height larger D_h **do**
10:            Split s_j horizontally in equidistant intervals D_h
11:     **for** Each segment s **do**
12:         $V = V \cup \{(x_m, y_m) \mid (x_m, y_m)$ is the centre of mass of segment $s\}$
13:     **for** Each pair of nodes $(u, v) \in V \times V$ **do**
14:         $E = E \cup (u, v)$ if the corresponding segments are connected in S
15:     **return** g

Fig. 4.10 shows examples of graph representations of the four different handwritten words in three different sizes, stemming from the four historical manuscripts. Note especially that certain (rather small) topological characteristics of the handwriting are not represented by small graphs (see, for example, Fig 4.10a and 4.10b). However, with larger graphs (i.e. smaller vertical and horizontal thresholds D_v and D_h, respectively) it can be observed that these characteristics are preserved in the graph representation. In Chapter 9 in Tables 9.3, 9.7, 9.11, and 9.15, further graph representations can be found for every manuscript.

Original	Preprocessed	Small	Medium	Large

(a) George Washington

(b) Parzival

(c) Alvermann Konzilsprotokolle

(d) Botany

Fig. 4.10: Original and preprocessed word images, as well as corresponding `Projection` graph representations stemming from different manuscripts.

4.3.4 *Split-Based Graphs*

The last graph representation is based on an adaptive and iterative segmentation of binarised word images B by means of horizontal and vertical splittings. Similar to `Projection`, the segmentation is based on projection profiles of word images. However, their algorithmic procedures are clearly distinguished from each other.

The procedure is formalised in Algorithm 4 (denoted by `Split` from now on). First, each segment is iteratively split into smaller subsegments until the width and height of all segments is below a certain threshold, D_w

and D_h, respectively. Initially B is regarded as one segment $\{s_1\}$ (see line 1 of Algorithm 4). Then, as long as there is a segment $s_i \in \{s_1, \ldots, s_n\}$ with a width greater than threshold D_w, this segment is vertically subdivided into subsegments $\{s_{i_1}, \ldots, s_{i_m}\}$ by means of the projection profile P_v of s_i (for details on this procedure we refer to Subsection 4.3.3). If the histogram P_v contains no white spaces, i.e. $p_i > 0$, $\forall\, p_i \in P_v$, the segment s_i is split in its vertical centre into two subsegments $\{s_{i_1}, s_{i_2}\}$ (see lines 3 to 8 of Algorithm 4 and Fig. 4.11a). Next, the same procedure as described above is applied to each segment with a height greater than threshold D_h in a horizontal, rather than vertical direction (see lines 9 to 13 of Algorithm 4 and Fig. 4.11b).

(a)   (b)

Fig. 4.11: Vertical and horizontal projection profile of the word "Orders". First, the binarised word image is segmented vertically at the white spaces in the projection profile in Subfigure (a). Next, segments are further segmented horizontally in Subfigure (b). If no white spaces are available, the segment is split in the centre.

If none of the segments can be split further, the centre of mass (x_m, y_m) is computed for each final segment s and a node is inserted into the graph labelled with the (x, y)-coordinates of the closest point on the skeletonised word image S to the centre of mass of this segment (x_m, y_m) (see lines 14 and 15 of Algorithm 4). If a segment consists of background pixels only, no node is created for this segment.

Finally, an undirected edge (u, v) between $u \in V$ and $v \in V$ is inserted into the graph for each pair of nodes, if the corresponding pair of segments is directly connected by a chain of foreground pixels in the skeletonised word image S (see lines 16 and 17 of Algorithm 4 and Fig. 4.9).

Fig. 4.12 shows examples of graph representations of the four different handwritten words in three different sizes, stemming from the four historical manuscripts. Note especially that even small graphs can represent the

---

Algorithm 4 Graph extraction based on splittings

In: Binary image B, Skeleton image S, Width threshold D_w, Height threshold D_h
Out: Graph $g = (V, E)$ with nodes V and edges E
1: **function** SPLIT(B,S,D_w,D_h)
2:    Add B to $\{s_1\}$
3:    **while** Segment $s_i \in \{s_1, \ldots, s_n\}$ has a width larger D_w or height larger D_h **do**
4:        **for** Each segment s_i with width larger D_w **do**
5:            **if** s_i contains white spaces in vertical projection profile P_v **then**
6:                Split s_i vertically at middle of white spaces of P_v into $\{s_{i_1}, \ldots, s_{i_m}\}$
7:            **else**
8:                Split s_i vertically at vertical centre of s_i into $\{s_{i_1}, s_{i_2}\}$
9:        **for** Each segment s_i with height larger D_h **do**
10:           **if** s_i contains white spaces in horizontal projection profile P_h **then**
11:               Split s_i horizontally at middle of white spaces of P_h into $\{s_{i_1}, \ldots, s_{i_m}\}$
12:           **else**
13:               Split s_i horizontally at horizontal centre of s_i into $\{s_{i_1}, s_{i_2}\}$
14:    **for** Each segment s **do**
15:        $V = V \cup \{(x_m, y_m) \mid (x_m, y_m)$ is the centre of mass of segment $s\}$
16:    **for** Each pair of nodes $(u, v) \in V \times V$ **do**
17:        $E = E \cup (u, v)$ if the corresponding segments are connected in S
18:    **return** g

---

handwriting characteristics quite well. In Chapter 9 in Tables 9.4, 9.8, 9.12, and 9.16, further graph representations can be found for every manuscript.

4.4 Preliminary Experimental Evaluation

We carry out a preliminary experimental evaluation by a distance-based *k-nearest neighbour* (k-NN) classification experiment on the four different manuscripts introduced in Chapter 3. Note that the k-NN evaluation cannot directly be compared with the prospective KWS experiments in Chapter 7. However, this examination will allow us to judge the quality of the introduced graph formalisms and the possibility of using them to represent handwritten manuscripts. In particular, we want to examine whether or not the proposed representations can be used with the more difficult KWS scenario.

A k-NN classifies an unknown object by its dissimilarity (i.e. distance) to its k-nearest neighbours in a training set. That is, an unknown object is assigned to the class of the majority of its k-nearest neighbours (e.g. for $k = 3$ with $\{a, b, a\}$ the unknown word is classified by a). In cases of a tie (i.e. when no majority is found), we employ $k = 1$.

Original	Preprocessed	Small	Medium	Large

(a) George Washington

(b) Parzival

(c) Alvermann Konzilsprotokolle

(d) Botany

Fig. 4.12: Original and preprocessed word images, as well as corresponding `Split` graph representations stemming from different manuscripts.

In this experiment, we employ a k-NN for the classification of handwritten word images that are represented by means of different graph formalisms. Hence, we require a graph dissimilarity measure for graphs. We make use of *graph edit distance* (GED), which measures the minimal amount of distortion needed to transform one graph into another (for details on GED and its computation we refer to Chapter 5).

For each manuscript, the parameters are optimised on ten different words with different word lengths. To this end, each word is classified in a leave-one-out setting by a validation set that consists of all or at least 10 random instances per word and a maximum of 900 additional random words (1,000 words in total). The optimised systems are then evaluated by twenty different words. That is, each word is classified by a test set that consists of all or at least 10 random instances per word and a maximum of 800 additional random words (similar to the validation set).

First, the cost functions of GED, as well as the parameters of the different graph representations, are optimised independently for each manuscript on the validation set. Next, the different graph representations are compared by means of k-NN classification using the optimised parameter values. That is, the accuracy for every graph representation (i.e. Keypoint, Grid, Projection, and Split) is measured for every $k \in \{1, \ldots, 20\}$. We compare all graph representations with each other on *George Washington* (GW) and *Parzival* (PAR), while on *Alvermann Konzilsprotokolle* (AK) and *Botany* (BOT), only the two representations that lead to the best results on GW/PAR are tested.

Note that in the current scenario, large values for k (i.e. $k > 15$) are better comparable with the KWS experiments where k is equal to the number of words in the complete document. Moreover, the classification becomes more difficult for large k, and thus, differences among the four graph representations will become more evident.

In Table 4.2, the classification accuracy for $k = 5, 10, 15, 20$ as well as the average over all k is shown in the GW dataset. We observe that all graph representations achieve high accuracy levels between 90% and 100% for $k = 5$. However, we can also observe that the accuracy level declines when k is increased. Overall, we observe the highest average accuracy on Keypoint, but Projection also achieves comparable results. The same observation can be made about Fig. 4.13, in which an accuracy plot is provided for all graph representations and all tested parameters k.

Table 4.2: GW: k-NN accuracy of different graph representations for $k = 5, 10, 15, 20$ as well as the average over all k.

Method	$k = 5$	$k = 10$	$k = 15$	$k = 20$	Average $k = \{1, \ldots, 20\}$	
Keypoint	100.0	95.0	95.0	85.0	93.50	(1)
Grid	90.0	80.0	70.0	75.0	83.25	(4)
Projection	95.0	90.0	85.0	80.0	89.50	(2)
Split	95.0	95.0	75.0	70.0	85.50	(3)

Generally, we observe lower accuracies on PAR when compared to the results on GW, as shown in Table 4.3. We see that Split and Projection achieve better results than Keypoint for $k = 5$. However, all graph representations are characterised by an accuracy drop for higher k. Overall, Keypoint achieves again the best average performance over all k. On the other hand, Grid again achieves rather low accuracy levels when compared to the remaining graph representations. In Fig. 4.14, the accuracy plot is provided for all graph representations over all k.

Comparing the average accuracies of all four graph representations of GW and PAR, we observe that Keypoint achieves the highest average of

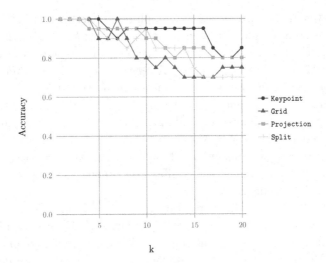

Fig. 4.13: GW: k-NN accuracy of different graph representations for $k = \{1, \ldots, 20\}$.

Table 4.3: PAR: k-NN accuracy of different graph representations for $k = 5, 10, 15, 20$ as well as the average over all k.

Method	$k = 5$	$k = 10$	$k = 15$	$k = 20$	Average $k = \{1, \ldots, 20\}$	
Keypoint	70.0	65.0	70.0	65.0	68.75	(1)
Grid	65.0	65.0	55.0	55.0	60.50	(4)
Projection	75.0	60.0	50.0	50.0	62.50	(3)
Split	80.0	55.0	60.0	55.0	66.00	(2)

81.13% over both manuscripts while Grid achieves the lowest average accuracy of 71.88%. Projection and Split achieve very similar averages of 76.00% and 75.75%, respectively. For further evaluations on AK and BOT, we consider Keypoint and Projection only.

In Table 4.4, the accuracy level for Keypoint and Projection is shown for the AK manuscript. For both graph representations we achieve very high accuracy levels for small $k = 5$ (no classification error), while a clear decline for both representations can be observed for higher k. Overall, Keypoint achieves slightly better accuracy when compared to Projection. The same observations can be made in the accuracy plot in Fig. 4.15.

Finally, we can observe very similar results on BOT for both graph rep-

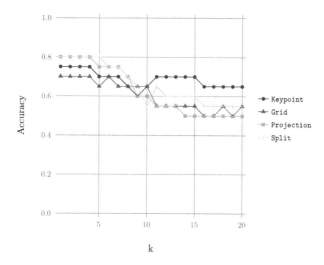

Fig. 4.14: PAR: k-NN accuracy of different graph representations for $k = \{1, \ldots, 20\}$.

Table 4.4: AK: k-NN accuracy of different graph representations for $k = 5, 10, 15, 20$ as well as the average over all k.

Method	$k = 5$	$k = 10$	$k = 15$	$k = 20$	Average $k = \{1, \ldots, 20\}$	
Keypoint	100.0	80.0	75.0	65.0	83.00	(1)
Projection	100.0	90.0	60.0	50.0	79.75	(2)

resentations, as shown in Table 4.5. While the Projection graphs achieve better results with $k = 5$, we see that Keypoint achieves higher accuracies with larger k. Overall, Keypoint with an average accuracy of 69.75% performs better than Projection, which achieves an accuracy level of 62.25%. In Fig. 4.16, this observation can also be made in the accuracy plot for all k.

As a final conclusion, we can say that Keypoint achieves the highest accuracy level over all manuscripts. Especially in the case of large numbers of neighbours, k, this graph representation generally achieves better results than the other representations. However, Projection and Split can also keep up with Keypoint in cases where only a small number of neighbours is considered.

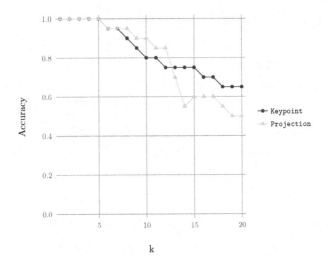

Fig. 4.15: AK: k-NN accuracy of different graph representations for $k = \{1, \ldots, 20\}$.

Table 4.5: BOT: k-NN accuracy of different graph representations for $k = 5, 10, 15, 20$ as well as the average over all k.

Method	$k = 5$	$k = 10$	$k = 15$	$k = 20$	Average $k = \{1, \ldots, 20\}$	
Keypoint	90.0	65.0	50.0	40.0	69.75	(1)
Projection	100.0	60.0	45.0	30.0	62.25	(2)

4.5   Summary

The vast majority of *keyword spotting* (KWS) approaches make use of a statistical representation of handwritten word images. That is, word images are represented by sequences of feature vectors, whereby different characteristics of handwriting are represented by means of ordered sets of numerical features. In contrast to this, only few KWS approaches are based on the representation of handwriting by means of structural approaches such as strings, trees, or graphs. This is rather surprising as structural approaches, and especially graph-based approaches, are characterised by their powerful and flexible representation possibilities. That is, graphs are able to adapt their structure and labelling to the complexity of the underlying pattern.

In this chapter, we propose four different formalisms for the representa-

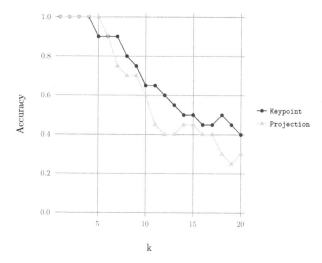

Fig. 4.16: BOT: k-NN accuracy of different graph representations for $k = \{1, \ldots, 20\}$.

tion of handwriting by graphs. The first representation formalism (denoted by Keypoint), is based on points with certain characteristics (so-called keypoints) such as the end and junction points of the handwriting. In particular, nodes represent these keypoints, while edges are inserted between keypoints that are connected by the handwriting strokes. The next graph representation (denoted by Grid), is based on a grid-wise segmentation of handwriting images. That is, a node represents the centre of a mass of a segment, while edges are constructed in a two-step procedure. First, edges are inserted between neighbouring segments, and second, the number of edges is reduced by means of a minimal spanning tree algorithm (actually transforming the graphs into trees). The remaining graph representations (denoted by Projection and Split) are based on an adaptive rather than static segmentation of the word images. That is, a word image is divided into smaller segments by means of vertical and horizontal projection profiles. Finally, nodes are used to represent the centre of mass of a segment, while edges are inserted between nodes that are connected by a chain of foreground pixels.

In a preliminary experimental evaluation, the four different graph representations are compared by means of a *k-nearest neighbour* (k-NN) classification experiment on four different historical manuscripts. It turns out that the graph representation Keypoint achieves the highest level of accuracy on all manuscripts. However, the graph representations `Projection` and `Split` also achieve high accuracy levels, especially with smaller parameter values for k. On the other hand, we observe that the graph representation `Grid` leads to the lowest accuracy on all four manuscripts. Overall, we can conclude that all representations are capable of capturing important handwriting characteristics. Hence, these graph representations might build the foundation for the graph-based KWS framework to be elaborated on this book.

Chapter 5

Graph Matching

5.1 Overview and Broader Perspective

In document analysis and related fields we have observed rather limited graph-based approaches until now [87, 195]. In particular, only few approaches have been proposed for *keyword spotting* (KWS) in handwritten historical documents [75–77]. To bridge this gap, we proposed several graph-based representation formalisms for handwriting in the previous chapter. These formalisms allow us to represent arbitrary handwritten word images by means of graphs. Consequently, (segmented) handwritten historical documents can be represented by a set of document graphs (i.e. one graph per word in the document). To make this set of graphs accessible for searching (for instance for keyword spotting), we need a method for the computation of pairwise dissimilarities or similarities between a given query graph and all document graphs.

The dissimilarity or similarity between two graphs is commonly measured by a specific *graph-matching* algorithm, as shown in the taxonomy in Fig. 5.1. In general, graph matching aims to map substructures of a first graph, g_1, to substructures of a second graph, g_2. On the resulting mapping, a dissimilarity score $d(g_1, g_2)$ (or a similarity score $s(g_1, g_2)$) can be derived that describes the proximity of the two graphs [196].

Basically, graphs can either be matched by means of *exact* or *inexact* graph matching approaches. Exact graph matching aims to find correspondence between parts or subparts of graphs which are identical in terms of their labels and their structure [196]. In contrast to this, inexact graph matching allows us to match pairs of graphs that have no common parts by endowing a certain error tolerance with respect to both structure and labelling.

Both *graph embedding* [197–199] and *graph kernel* [200–202] aim to bridge the gap between structural and statistical approaches by means of explicit or implicit projections of the underlying graphs into a real vector space or Hilbert space [183]. Eventually, arbitrary statistical distances and/or algorithms developed for feature vectors can be employed in the embedding space. For this embedding, both approaches make use of some graph dissimilarity or similarity measures.

In this book we focus on the application of graph matching approaches in the domain of KWS. This chapter provides an introduction to graph matching and related concepts. For reason of self-containment, we review known graph matching concepts [79, 93–95] and provide a thorough and consistent summary of them. In particular, we follow and summarise certain argumentations on formulations as provided in [183][1]. Note that this book is not claiming any intellectual contribution with respect to the employed graph matching algorithms as such. Moreover, cross-references are clearly marked as such throughout. For a more thorough review on graph-based methods and applications, the reader is referred to [189, 203, 204].

In Section 5.2, we review both exact and inexact graph matching with respect to their applicability to KWS. Next, in Section 5.3 we review one specific inexact graph matching concept that can be employed in the case of graph-based KWS, namely *graph edit distance* (GED). Finally, a novel inexact graph matching algorithm by means of polar segmentations is introduced in Section 5.4.

5.2   Graph Matching for KWS

In this subsection we give a formal definition of both exact and inexact graph matching. Based on this formal introduction, we review the intrinsic properties of both approaches with respect to the requirements of KWS. Hence, we evaluate whether either of the approaches can be employed in this specific use-case.

5.2.1   *Exact Graph Matching*

Exact graph matching ensures complete mapping between two graphs, g_1 and g_2, that violates neither the edge structures nor the labelling of both graphs. This mapping is commonly referred to as *graph isomorphism*. That

---

[1]Sections 5.2.1, 5.2.2, and 5.3 of this book follow Sections 1.3.1, 1.3.2, and 2.1 of [183].

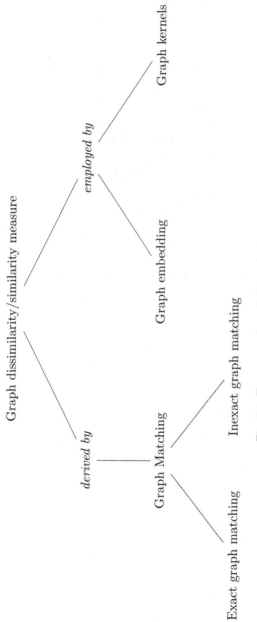

Fig. 5.1: Taxonomy of graph dissimilarity measures.

is, graph isomorphism ensures a bijective mapping of the nodes of g_1 to the nodes of g_2 with identical labels, while keeping the structural relationships given by the edges of g_1 and g_2. That is, the mapping ensures that the nodes of g_1 and g_2 are only mapped if there is an edge between the corresponding nodes in g_1 and in g_2 [183].

In contract to this, *subgraph isomorphism* verifies whether or not a smaller graph is completely available within a larger graph. That is, subgraph isomorphism ensures an injective mapping between g_1 and an induced subgraph of g_2 [183]. Due to its weaker mapping condition, subgraph isomorphism is known to be $\mathcal{NP}$-*complete* [205] (while for graph isomorphism the complexity is — most probably — quasipolynomial [206]).

Fig. 5.2 provides examples of graph and subgraph isomorphisms, as well as corresponding mapping functions $f : V_1 \rightarrow V_2$ (i.e. the bijective function for graph isomorphism) and $f : V_2 \rightarrow V_3$ (i.e. the injective function for subgraph isomorphism). Note that the nodes are labelled in four different colours, while the edges remain unlabelled. For better readability, the nodes are enumerated.

Common approaches for the computation of graph isomorphism are based on tree search procedures with backtracking [183]. To limit the exponential search space, most approaches make use of certain heuristics [207–215]. Moreover, note that there are a number of approaches that are not based on tree search algorithms, such as, for example, *Nauty* [216, 217], *random walks* [218, 219] and *decision trees* [220, 221], to name just a few alternatives. However, with respect to a graph dissimilarity (or similarity) measure, graph isomorphism and subgraph isomorphism only allow us to determine whether two graphs or at least parts of the graphs are identical in terms of labelling and structure. As a result, we can derive a binary graph dissimilarity measure only, i.e. $d(g_1, g_2) = 1$ for non-isomorphic graphs or graphs for which there is no subgraph isomorphism existing, and $d(g_1, g_2) = 0$ for isomorphic graphs or graphs for which there is a subgraph isomorphism existing [183].

More refined dissimilarity measures have been proposed on the basis of graph isomorphism [222]. In particular, these measures make use of the minimum and maximum common substructures of two graphs, i.e. the *minimum common supergraph* (MCS) [223] and the *maximum common subgraph* (mcs) [224, 225].

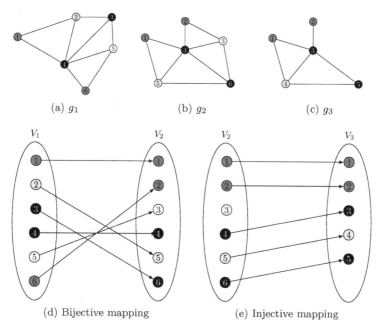

(a) g_1 (b) g_2 (c) g_3

(d) Bijective mapping   (e) Injective mapping

Fig. 5.2: Graph g_1 (a) and graph g_2 (b) are isomorphic. Graph g_3 (c) is a subgraph of graphs g_1 and g_2. Bijective mapping of g_1 to g_2 (d), as well as injective mapping of g_2 to g_3 (e).

5.2.2   *Inexact Graph Matching*

In contrast to exact graph matching, inexact graph matching permits mapping between two graphs, g_1 and g_2, that have no parts in common by endowing a certain error-tolerance with respect to both structure and labelling. In particular, inexact graph matching relaxes the concept of graph isomorphism by allowing the mapping of nodes and edges with different labels, by allowing mappings that possibly violate the edge structure, and by allowing deletions and/or insertions of nodes and edges [183].

Fig. 5.3 shows an error-tolerant graph matching between graph g_1 and graph g_2 by means of a bijective mapping between V_1 and V_2. With ε we denote an empty node. Note that the nodes are labelled in four different colours, while the edges remain unlabelled. For better readability, the nodes are enumerated from top left to bottom right.

We observe a substitution of two nodes with different labels (node (2) in g_1 is substituted for node (1) in g_2). Moreover, we observe mapping that violates the edge structure (e.g. (3) $\Rightarrow$ (2) and (5) $\Rightarrow$ (3)). Finally, we observe a node insertion, as well as a node deletion (e.g. (ε) $\Rightarrow$ (7) and (7) $\Rightarrow$ (ε)).

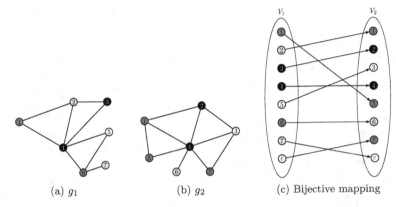

(a) g_1 (b) g_2 (c) Bijective mapping

Fig. 5.3: Error-tolerant graph matching of graph g_1 (a) and graph g_2 (b) by means of bijective mapping between $V_1 \cup \{\varepsilon\}$ and $V_2 \cup \{\varepsilon\}$ (c). Symbol ε refers to empty nodes.

As described in [183], the matching of edges can actually always be derived from the matching of nodes. That is, given a node, matching edges are matched with respect to the matching of the adjacent nodes. In particular, [183] distinguishes between three different cases, namely mapping, deletion, and insertion of edges, as shown in Fig. 5.4.

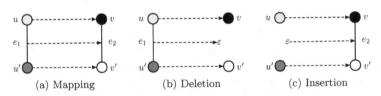

(a) Mapping   (b) Deletion   (c) Insertion

Fig. 5.4: Three different matching cases of edges e_1 and e_2 shown by means of adjacent nodes u, u' and v, v', respectively.

Each node and edge matching operation (i.e. mapping, insertion, and deletion) can be associated with a certain cost function c. As a result, inexact graph matching algorithms aim to minimise the overcall cost $c(f)$ given by the mapping $f(g_1, g_2)$ of graph g_1 to g_2, i.e. by the sum of all matching operations given by all node and edge mappings, deletions, and insertions [183].

The minimisation of $c(f)$ is known to be $\mathcal{NP}$-*complete*, and thus, several fast approximations have been proposed in recent decades [78, 79, 226–232]. Similarly to approaches for the computation of graph isomorphism, several methods based on tree search algorithms have been proposed for inexact graph matching [233–235]. Approaches based on tabu search procedures have also been proposed [236, 237]. Another popular approach is to re-formulate the discrete graph matching problem as a continuous optimi-sation problem that can then be solved using various optimisation meth-ods [238–241]. Spectral methods make use of eigenvectors of an adjacency or *Laplacian* matrix of graphs, which is invariant with respect to node per-mutations [242–248]. Finally, graph kernels that can be seen as both a graph dissimilarity measure and an embedding of graphs in a Hilbert space have been proposed [183, 202]. Popular graph kernels are, for example, *ran-dom walk kernels* [249–253], *convolution kernels* [254–259], and *diffusion kernels* [260–262], to name just three examples.

5.2.3 *Formal Requirements*

Given the characteristic of exact and inexact graph matching, the ques-tions arises as to whether one of the two approaches is feasible for KWS. Exact graph matching finds correspondences between identical parts or sub-parts of graphs, while inexact graph matching permits matching between non-identical parts or subparts of graphs. In the case of KWS, graphs are used to represent single word images, as shown in Fig. 5.5. However, these word images have intraword variations with respect to both writing style (e.g. skew, slant and scale) and the document itself (e.g. noise, signs of degradation, ink bleed-through), as shown in Fig 5.5a. As a result, not only are the preprocessed word images affected by such intraword variations but their corresponding graph representations are also affected, as shown in Figs. 5.5b and 5.5c, respectively. Hence, we can conclude that exact graph matching is not well-suited to handwritten document analysis applica-tions. As a consequence, this book focuses on inexact graph matching from now on.

In the case of template-based KWS, another important fact has to be considered, namely the vast number of comparisons needed to perform a particular retrieval request. That is, given a document with N words and n different queries, one needs to perform $N \times n$ pairwise matchings between

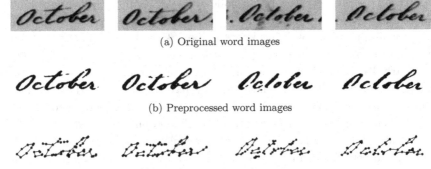

(a) Original word images

(b) Preprocessed word images

(c) Keypoint graph representation

Fig. 5.5: Intraword variations of the word "October" in the original word images in Subfigure (a), the preprocessed word images in Subfigure (b), and the graph representations in Subfigure (c).

query and document words. In the case of a large document with, for example, 100,000 different words and 100 queries, one would need to perform 10,000,000 comparisons. Every word image is often represented by a graph with about 75 to 100 nodes and edges, and thus, an optimal error-tolerant graph matching quickly becomes intractable due to its $\mathcal{NP}$-completeness. As a consequence, only suboptimal procedures for inexact graph matching that can match graphs in polynomial rather than exponential time can be employed. In the following, we focus on approaches for suboptimal inexact graph matching with respect to two different paradigms, namely the well-known *graph edit distance* (GED) [263, 264] and a novel approach called *polar graph dissimilarity* (PGD) in Sections 5.3 and 5.4, respectively.

5.3 Graph Edit Distance

The content of this section describes basic definitions, formulations and methods of the concept of *graph edit distance* (GED). GED is one of the basic building blocks of our research on graph-based KWS and is included here for the sake of the self-containment of this book. Note that we follow the argumentation and formulations of [79, 183] to a large extent. That is, the reformulation of GED as a *bipartite matching problem* as well as the use of *Hausdorff* distance models for graph matching should not be considered as intellectual contributions in this book.

The concept of GED is considered to be one of the most flexible and versatile graph matching models available [183, 189, 203]. GED allows us to embed specific domain knowledge to the matching by means of user-defined cost functions. However, the major drawback of GED is its computational complexity, which restricts its applicability to graphs of rather small size [183]. In fact, GED belongs to the family of *quadratic assignment problems* (QAPs) [265], which in turn belong to the class of $\mathcal{NP}$-complete problems[2]. Basically, QAPs deal with the optimal assignment of entities of two sets. To overcome the large computational complexity of GED, we focus on suboptimal approaches for GED with cubic and quadratic complexity from now on.

Originally, the concept of *edit distance* (also known as *Levenshtein* distance) was proposed for strings [266]. String edit distances measure the minimum amount of distortion needed to transform one string into another by using a set of different edit operations (for instance the *insertion, deletion,* or *substitution* of symbols). The concept of edit distance was adapted for trees [267] and later for graphs [263, 264, 268–270].

Similar to string edit distance, the basic idea of graph edit distance is to transform a source graph $g_1 = (V_1, E_1, \mu_1, \nu_1)$ into a target graph $g_2 = (V_2, E_2, \mu_2, \nu_2)$ using a set of edit operations applicable on both nodes and edges [183]. Typical edit operations are the substitution, deletion, and insertion of both nodes and edges. Further edit operations such as *merging* and *splitting* have been proposed [271] but are not employed in this book. Formally, we define the following node edit operations for g_1 and g_2

(1) Node substitution $u \in V_1$ and $v \in V_2$ denoted by $(u \to v)$,
(2) Node deletion $u \in V_1$ denoted by $(u \to \varepsilon)$,
(3) Node insertion $v \in V_2$ denoted by $(\varepsilon \to v)$,

where ε refers to the empty node. For edge edit operations, we use a similar notation.

To transform one graph into another, a set of k edit operations $\{e_1, \ldots, e_k\}$ is required (unless g_1 is isomorphic to g_2). A complete set of edit operations that transforms g_1 into g_2 is called a *(complete) edit path* [183]. A *partial edit path*, i.e. a subset of $\{e_1, \ldots, e_k\}$, transforms subsets of nodes and/or edges of the underlying graphs. In both cases, edge edit operations are implicitly given by the corresponding node edit operations in an edit path [183].

---

[2]As stated in [183], an exact and efficient algorithm for the graph edit distance problem cannot be developed unless $\mathcal{P} = \mathcal{NP}$.

In Fig. 5.6, two exemplary edit paths $\lambda_1(g_1, g_2)$ and $\lambda_2(g_1, g_2)$ of two undirected graphs g_1 and g_2 are shown. In Fig. 5.6a, the edit path is given by

$$\lambda = \{(u_1 \to \varepsilon), (u_2 \to v_1), (u_4 \to v_3), (u_3 \to v_2)\} \quad ,$$

while edges are implicitly edited by

$$\{((u_1, u_3) \to \varepsilon), ((u_1, u_4) \to \varepsilon), ((u_2, u_4) \to (v_1, v_3)),$$
$$((u_2, u_3) \to (v_1, v_2)), (\varepsilon \to (v_2, v_2))\} \quad ,$$

according to the given node edit operations.

In Fig. 5.6b, the edit path is given by

$$\lambda = \{(u_4 \to v_3), (u_3 \to v_2), (u_2 \to v_1), (u_1 \to \varepsilon)\} \quad ,$$

while edges are implicitly edited by

$$\{(\varepsilon \to (v_2, v_3)), ((u_2, u_4) \to (v_1, v_3)), ((u_2, u_3) \to (v_1, v_2)),$$
$$((u_1, u_3) \to \varepsilon), ((u_1, u_4) \to \varepsilon)\} \quad ,$$

according to the given node edit operations.

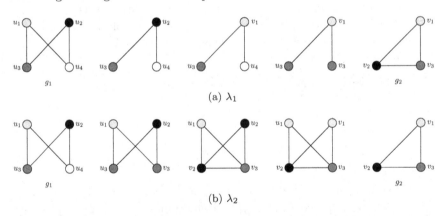

(a) λ_1

(b) λ_2

Fig. 5.6: Two exemplary edit paths $\lambda_1(g_1, g_2)$ and $\lambda_2(g_1, g_2)$ that transform graph g_1 into g_2 and vice versa.

For every pair of graphs g_1 and g_2 in general, different edit paths $\Upsilon(g_1, g_2)$ exist, as illustrated above. Basically, every edit operation e is assigned a cost $c(e)$. In general, edit costs $c(e)$ should reflect the degree of deformation e, i.e. strong modifications of the graph should lead to high edit costs and, vice versa; smaller or negligible modifications should lead to smaller edit costs [183]. Moreover, edit costs can be used to reflect specific domain knowledge.

The *graph edit distance* (GED) $d_{\lambda_{min}}(g_1, g_2)$, or $d_{\lambda_{min}}$ for short, between a source graph $g_1 = (V_1, E_1, \mu, \nu)$ and a target graph $g_2 = (V_2, E_2, \mu, \nu)$ is defined by [183]

$$d_{\lambda_{min}}(g_1, g_2) = \min_{\lambda \in \Upsilon(g_1, g_2)} \sum_{e_i \in \lambda} c(e_i) \quad , \tag{5.1}$$

where $\Upsilon(g_1, g_2)$ denotes the set of all complete edit paths transforming g_1 into g_2, c denotes the cost function measuring the strength $c(e_i)$ of node edit operation e_i (including the cost of all edge edit operations implied by the operations applied to the adjacent nodes of the edges), and λ_{min} refers to the minimal cost edit path found in $\Upsilon(g_1, g_2)$ [183].

5.3.1 Suboptimal Algorithms

The exact computation of graph edit distance is exponential in the number of nodes of the involved graphs [183], and thus not applicable in the case of large numbers of matchings and/or large graphs. In order to increase the applicability of graphs, several fast but suboptimal algorithms for GED have been proposed [183,189,203]. In the following, several of these methods are summarised.

The *bipartite* graph-matching algorithm introduced in [93], for instance, reduces the QAP of graph edit distance computation to an instance of a *linear sum assignment problem* (LSAP). Similarly to QAPs, LSAPs deal with the question of how the entities of two sets can be optimally assigned to each other [272]. However, a large number of efficient algorithms exist to solve LSAPs (see [273] for an exhaustive survey). The time complexity of the best performing exact algorithms for LSAPs is cubic in the size of the problem. Hence – in contrast with QAPs – LSAPs can be solved quite efficiently.

In [240], GED is represented as a *binary linear programming* (BLP) formulation that is an instance of an LSAP. However, BLP is limited to undirected graphs with labelled nodes and unlabelled edges, and thus, lacks structural assignment information. To overcome this limitation, a reformulation for BLP is proposed in [274].

Rather than reducing GED to an instance of LSAP, different approximations for QAP are proposed in [275]. Last but not least, the problem of GED can be suboptimally computed by means of a *Hausdorff* set matching that can in turn be solved in quadratic time complexity [79].

In [276, 277], a depth-first search *anytime* algorithm for GED is proposed. That is, the approximation error with respect to the effective GED becomes smaller by the time of execution. In particular, a valid but suboptimal GED is provided at any point in the assignment process. Note that all other suboptimal algorithms for GED discussed above lead to a single GED.

In this book, we make use of four different approximations for GED, namely *bipartite graph edit distance* (BP) [93], *Hausdorff edit distance* (HED) [79], *context-aware Hausdorff edit distance* (CED) [94], and *bipartite graph edit distance 2* (BP2) [95]. In the following four subsections, we provide a summary of each of the four methods.

5.3.2 Bipartite Graph Edit Distance (BP)

In this subsection, we review an algorithm that is able to return an upper bound for graph edit distance, termed *bipartite graph edit distance* (BP) [93]. Basically, BP optimally solves an LSAP which is stated on assignments of local structures to the involved graphs (namely nodes and adjacent edges)[3]. This assignment of local structures can then be used to infer a complete set of globally consistent node and edge edit operations, i.e. a valid edit path $\lambda \in \Upsilon(g_1, g_2)$. The sum of the costs of this – not necessarily optimal – edit path gives us an upper bound on the exact distance $d_{\lambda_{\min}}$ [278]. In recent years, BP has been extended in various ways [196, 277–280]. In particular, we observe improvements with respect to runtime [173, 230, 280, 281] and matching accuracy [196, 272, 279, 282–292].

In order to reformulate the graph edit distance problem to an instance of an LSAP, the use of a square cost matrix $\mathbf{C}$ is proposed in [93]. Formally, based on the node sets $V_1 = \{u_1, \ldots, u_n\}$ and $V_2 = \{v_1, \ldots, v_m\}$ of g_1 and g_2, respectively, a cost matrix $\mathbf{C}$ is defined by

$$\mathbf{C} = \begin{bmatrix} c_{11} & c_{12} & \cdots & c_{1m} & c_{1\varepsilon} & \infty & \cdots & \infty \\ c_{21} & c_{22} & \cdots & c_{2m} & \infty & c_{2\varepsilon} & \ddots & \vdots \\ \vdots & \vdots & \ddots & \vdots & \vdots & \ddots & \ddots & \infty \\ c_{n1} & c_{n2} & \cdots & c_{nm} & \infty & \cdots & \infty & c_{n\varepsilon} \\ c_{\varepsilon 1} & \infty & \cdots & \infty & 0 & 0 & \cdots & 0 \\ \infty & c_{\varepsilon 2} & \ddots & \vdots & 0 & 0 & \ddots & \vdots \\ \vdots & \ddots & \ddots & \infty & \vdots & \ddots & \ddots & 0 \\ \infty & \cdots & \infty & c_{\varepsilon m} & 0 & \cdots & 0 & 0 \end{bmatrix},$$

---

[3]LSAP is also termed *bipartite matching problem*.

where entry c_{ij} denotes the cost of a node substitution $(u_i \rightarrow v_j)$, $c_{i\varepsilon}$ denotes the cost of a node deletion $(u_i \rightarrow \varepsilon)$, and $c_{\varepsilon j}$ denotes the cost of a node insertion $(\varepsilon \rightarrow v_j)$.

Based on C, the following LSAP could be solved [93]

$$(\varphi_1^*, \ldots, \varphi_{(n+m)}^*) = \underset{(\varphi_1^*, \ldots, \varphi_{(n+m)}^*) \in \mathcal{S}_{(n+m)}}{\arg\min} \sum_{i=1}^{n+m} c_{i\varphi_i^*} \quad .$$

However, as stated in [183], this term omits the quadratic term during the assignment process, and thus we ignore the structural relationships between the nodes (i.e. the edges between the nodes). In order to integrate knowledge about the graph structure to each entry $c_{ij} \in \mathbf{C}$, i.e. to each cost of a node edit operation $(u_i \rightarrow v_j)$, it is proposed in [183] to add the minimum sum of edge edit operation costs, implied by the corresponding node operation. Formally, for every entry c_{ij} in the cost matrix $\mathbf{C}$ we solve an LSAP on the in- and outgoing edges of node u_i and v_j and add the resulting cost to c_{ij} [183]. That is, one can define

$$c_{ij}^* = c_{ij} + \underset{(\varphi_1, \ldots, \varphi_{(n+m)}) \in \mathcal{S}_{(n+m)}}{\min} \sum_{k=1}^{n+m} (c(a_{ik} \rightarrow b_{j\varphi_k}) + c(a_{ki} \rightarrow b_{\varphi_k j})) \quad ,$$

where $\mathcal{S}_{(n+m)}$ refers to the set of all $(n+m)!$ possible permutations of the integers $1, \ldots, (n+m)$.

The cost of the deletion of all adjacent edges of u_i can be added to entry $c_{i\varepsilon}$, which denotes the cost of a node deletion, and the cost of all insertions of the adjacent edges of v_j can be added to the entry $c_{\varepsilon j}$, which denotes the cost of a node insertion [183]. The cost matrix which is enriched with local structural matching information is denoted by $\mathbf{C}^* = (c_{ij}^*)$ in [183] (we use the same formalism).

This particular encoding of the minimum matching cost arising from the local edge structure enables the LSAP to consider information about the local, but not global, edge structure of a graph [183]. Note that in [282,286, 287] other approaches to encoding the local edge structure into individual entries $c_{ij}^* \in \mathbf{C}^*$ are proposed (e.g. by embedding a larger node context than adjacent edges only).

A minimum cost permutation $(\varphi_1^*, \ldots, \varphi_{n+m}^*)$ can now be derived on $\mathbf{C}^* = (c_{ij})$ via the LSAP solving algorithm (for instance by means of the *Munkres* algorithm [293] in [93], or the *Volgenant-Jonker* [294] in [281]).

This permutation corresponds to a bijective node assignment [93]

$$\psi = ((u_1 \rightarrow v_{\varphi_1}), (u_2 \rightarrow v_{\varphi_2}), \dots, (u_{m+n} \rightarrow v_{\varphi_{m+n}}))$$

of all nodes of g_1 to all nodes of g_2. Assignment ψ includes edit operations of the form $(u_i \rightarrow v_j)$, $(u_i \rightarrow \varepsilon)$, and $(\varepsilon \rightarrow v_j)^4$. Hence as stated in [183], LSAP finds a permutation which refers to an admissible and complete (but not necessarily optimal) edit path between the graphs under consideration, i.e. $\psi \in \Upsilon(g_1, g_2)$.

In Fig. 5.7, an example of a bijective node assignment between two graphs g_1 and g_2 is shown.

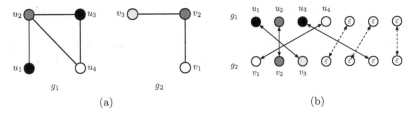

Fig. 5.7: Bijective node assignment (see (b)) of two graphs (see (a)) given by BP.

Remember that edge edit operations are implicitly given by the edit operations on their adjacent nodes. Hence, we can use the node assignment ψ, or more precisely, the minimum cost permutation $(\varphi_1^*, \dots, \varphi_{n+m}^*)$ derived on $\mathbf{C}^*$, to infer globally consistent edge edit operations [93]. The sum of costs of the node edit operations plus the sum of costs of the implied edge edit operations leads to an upper bound GED^5. Formally,

$$\mathbf{A} = \begin{array}{c} \\ 1 \\ \vdots \\ n \\ 1 \\ \vdots \\ m \end{array} \begin{bmatrix} \begin{array}{ccc} 1 & \dots & n \end{array} & \begin{array}{ccc} 1 & \dots & m \end{array} \\ \begin{array}{ccc} a_{11} & \dots & a_{1n} \\ \vdots & \ddots & \vdots \\ a_{n1} & \dots & a_{nn} \end{array} & \begin{array}{ccc} \varepsilon & \dots & \varepsilon \\ \vdots & \ddots & \vdots \\ \varepsilon & \dots & \varepsilon \end{array} \\ \begin{array}{ccc} \varepsilon & \dots & \varepsilon \\ \vdots & \ddots & \vdots \\ \varepsilon & \dots & \varepsilon \end{array} & \begin{array}{ccc} \varepsilon & \dots & \varepsilon \\ \vdots & \ddots & \vdots \\ \varepsilon & \dots & \varepsilon \end{array} \end{bmatrix} \quad (5.2)$$

---

^{4}Edit operations of the form $(\varepsilon \rightarrow \varepsilon)$ can be dismissed.
^{5}See [183] for proofs on this upper bound property.

and

$$
\mathbf{B} = \begin{array}{c}
\begin{array}{ccccccc} 1 & \dots & m & 1 & \dots & n \end{array} \\
\begin{array}{c} 1 \\ \vdots \\ m \\ 1 \\ \vdots \\ n \end{array}
\left[\begin{array}{ccc|ccc}
b_{11} & \dots & b_{1m} & \varepsilon & \dots & \varepsilon \\
\vdots & \ddots & \vdots & \vdots & \ddots & \vdots \\
b_{m1} & \dots & b_{mm} & \varepsilon & \dots & \varepsilon \\
\varepsilon & \dots & \varepsilon & \varepsilon & \dots & \varepsilon \\
\vdots & \ddots & \vdots & \vdots & \ddots & \vdots \\
\varepsilon & \dots & \varepsilon & \varepsilon & \dots & \varepsilon
\end{array} \right]
\end{array}
\qquad (5.3)
$$

are the adjacency matrices of graph g_1 and g_2, used to infer BP in [93]

$$
BP(g_1, g_2) = d_\psi = \sum_{i=1}^{n+m} c_{i\varphi_i^*} + \sum_{i=1}^{n+m} \sum_{j=1}^{n+m} c(a_{ij} \to b_{\varphi_i^* \varphi_j^*}) \quad . \qquad (5.4)
$$

The linear term of Eq. 5.4 refers to the cost of node edit operations captured in ψ (but this time applied to $\mathbf{C}$ rather than to $\mathbf{C}^*$). The quadratic term refers to the globally consistent edge edit operations that are implied by the node edit operations in ψ. In the case of undirected graphs, the quadratic term has to be multiplied by $\frac{1}{2}$ (as stated in [183]) or the inner sum has to be adapted such that to $j = i+1, \dots, n+m$ (as stated in [183]).

5.3.3 *Hausdorff Edit Distance (HED)*

In this subsection, we review a lower bound approximation for GED, namely *Hausdorff edit distance* (HED) with quadratic — rather than cubic — time complexity [79]. HED reformulates the graph edit distance problem as a set matching between local substructures. In this matching the assignments are independent of each other. In particular, each node of a source graph is compared with each node of a target graph similar to the Hausdorff distance between finite sets [295].

The Hausdorff distance H, as proposed in [295], is sensitive to outliers, and thus, a modified Hausdorff distance H' based on the sum — rather than the maximum — among all nearest neighbour distances is proposed [296]. This is particularly important in the case of KWS, where, for example, outliers appear based on flourish, as well as on the naturally high variability of handwriting (see Fig. 5.5 for an illustrative example). Moreover, outliers can also be caused by certain image preprocessing steps, such as, for example, imperfect binarisation or skeletonisation. Hence, the Hausdorff distance H would be largely influenced by such outliers in the handwriting.

Formally, given two sets of entities $A = \{a_1, \ldots, a_n\}$ and $B = \{b_1, \ldots, b_n\}$ the modified Hausdorff distance H' is defined by [296]

$$H'(A, B) = \sum_{a \in A} \min_{b \in B} d(a, b) + \sum_{b \in B} \min_{a \in A} d(a, b) \quad .$$

The HED for two graphs $g_1 = (V_1, E_1, \mu_1, \nu_1)$ and $g_2 = (V_2, E_2, \mu_2, \nu_2)$ can be derived from this particular distance measure by [79]

$$HED(g_1, g_2) = \sum_{u \in V_1} \min_{v \in V_2 \cup \{\epsilon\}} f(u, v) + \sum_{v \in V_2} \min_{u \in V_1 \cup \{\epsilon\}} f(u, v) \quad ,$$

where $f(u, v)$ is a cost function for node matchings. Formally $f(u, v)$ is defined as [79]

$$f(u, v) = \begin{cases} c(u \to \varepsilon) + \sum_{i=1}^{|P|} \frac{c(p \to \varepsilon)}{2} & \text{for node deletions } (u \to \epsilon) \\ c(\varepsilon \to u) + \sum_{i=1}^{|Q|} \frac{c(\varepsilon \to q)}{2} & \text{for node insertions } (\epsilon \to v) \\ \frac{c(u \to v) + \frac{HED(P,Q)}{2}}{2} & \text{for node substitutions } (u \to v) \end{cases} \quad , \tag{5.5}$$

where $P = \{p_1, \ldots, p_{|P|}\}$ is the set of edges adjacent to u and $Q = \{q_1, \ldots, q_{|Q|}\}$ is the set of edges adjacent to v. In contrast to BP, HED does not enforce bidirectional assignments for the involved nodes (see Fig. 5.8 for an illustrative example), and thus, substitution costs are distributed over two summation terms. In particular, HED allows directed and multiple node assignments. That is, each node (of both graphs) is potentially substituted twice and thus the substitution cost has to be divided by 2 [79]. Node deletions and insertions, on the other hand, only appear in one of the summation terms, and thus, their full cost is taken into account in $f(u, v)$ [79].

Similar to BP, edge edit costs are implicitly given by the node substitutions in HED. The set of edges P and Q, adjacent to the nodes actually matched, are assigned to all others by means of Hausdorff edit distance (similar to the node sets) [79]. Formally,

$$HED(P, Q) = \sum_{p \in P} \min_{q \in Q \cup \{\epsilon\}} g(p, q) + \sum_{q \in Q} \min_{p \in P \cup \{\epsilon\}} g(p, q) \quad ,$$

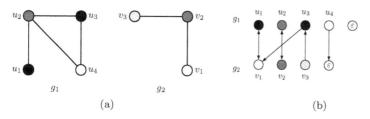

Fig. 5.8: Multiple and directed node assignments (b) of two graphs (a) given by HED.

where $g(p, q)$ is a cost function for edge matchings, formally given by [79]

$$g(p,q) = \begin{cases} c(p \to \varepsilon) & \text{for edge deletions } (p \to \epsilon) \\ c(\varepsilon \to q) & \text{for edge insertions } (\epsilon \to q) \\ \frac{c(p \to q)}{2} & \text{for edge substitutions } (p \to q) \end{cases} .$$

To ensure the lower bound property of HED, all node and edge edit costs in $f(u, v)$ and $g(p, q)$ are divided by 2 [79]. In order to limit the lower bound of HED, a minimum edit cost for $HED(g_1, g_2)$ and $HED(P, Q)$ can be used [79]. Formally,

$$HED(g_1, g_2) = \begin{cases} (|V_1| - |V_2|) \cdot \min_{u \in V_1} c(u \to \epsilon), & \text{if } |V_1| > |V_2| \\ (|V_2| - |V_1|) \cdot \min_{v \in V_2} c(\epsilon \to v), & \text{otherwise} \end{cases}$$

$$HED(P, Q) = \begin{cases} (|P| - |Q|) \cdot \min_{p \in P} c(p \to \epsilon), & \text{if } |P| > |Q| \\ (|Q| - |P|) \cdot \min_{q \in Q} c(\epsilon \to q), & \text{otherwise} \end{cases}$$

Hence, these lower bounds assert a minimum amount of deletion and insertion costs if the two matched sets differ in size [79].

5.3.3.1 *Context-aware Hausdorff Edit Distance (CED)*

Context-aware Hausdorff edit distance (CED) [94] is a direct extension of HED [79] with respect to the basic assignment procedure. That is, CED integrates a larger local context — rather than adjacent edges only — during the optimisation of the assignment. This allows us to better integrate the structural context into the set matching process, and thus, might reduce the approximation error with respect to the exact GED[6].

---

[6]The lower bound property of HED, however, is lost with CED.

Given a graph g, a structural node context of node u with degree n is defined as $c_n(u,g) = (L_1, \ldots, L_n)$, where L_i refers to the subgraphs containing all nodes (including the edges) that have a shortest path from u with a distance of $(i-1)$ $(i = 1, \ldots, n)$ [94]. In Fig. 5.9, an illustrative example of the structural node context is given.

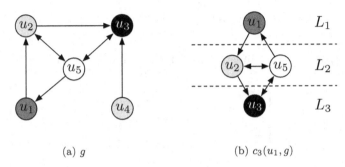

(a) g (b) $c_3(u_1, g)$

Fig. 5.9: In Subfigure (b) the structural node context of node u_1 with degree $n = 3$ of graph g in Subfigure (a) is shown.

To integrate the structural node context into the Hausdorff edit distance, a distance between two node contexts $c_n(u, g_1) = (L_1, \ldots, L_n)$ and $c_n(v, g_2) = (M_1, \ldots, M_n)$ has first to be defined. Formally, the distance between these two contexts is given by [94]

$$d(c_n(u, g_1), c_n(v, g_2)) = \sum_{i=1}^{n} HED(L_i, M_i) \quad .$$

Based on this distance, the nearest neighbour node $b_n(u, g_2) \in V_2$ of $u \in V_1$ and the nearest neighbour node $b_n(v, g_1) \in V_1$ of $v \in V_2$ is found by [94]

$$b_n(u, g_2) = \underset{v \in V_2}{\operatorname{argmin}} \, d(c_n(u, g_1), c_n(v, g_2))$$
$$b_n(v, g_1) = \underset{u \in V_1}{\operatorname{argmin}} \, d(c_n(u, g_1), c_n(v, g_2))$$

That is, $b_n(u, g_2) \in V_2$ and $b_n(v, g_1) \in V_1$ are the most similar nodes of $u \in V_1$ and $v \in V_2$, respectively (using a node context of level n). Lastly, the Hausdorff edit distance embeds these nearest neighbours by [94]

$$CED(g_1, g_2, n) = \sum_{u \in V_1} \min_{v \in \{b_n(u,g_2),\varepsilon\}} f(u,v) + \sum_{v \in V_2} \min_{u \in \{b_n(v,g_1),\varepsilon\}} f(u,v) \quad .$$

Hence, every node is either assigned to its nearest neighbour in the other graph or deleted/inserted. Note that CED corresponds to the original HED in the case of $n = 1$ (i.e. considering only nodes and adjacent edges as local substructures).

5.3.4 *Bipartite Graph Edit Distance 2 (BP2)*

In this subsection we review another approximation algorithm for GED, namely the *bipartite graph edit distance 2* (BP2) [95]. BP2 can be seen as a combination of the two previous approaches, namely HED and a greedy version of BP. In particular, BP2 combines the upper bound property of BP with the quadratic time complexity of HED.

Similar to HED, the node assignment strategy is based on bipartite graphs as illustrated in an example in Fig. 5.10. However, in contrast to HED, bijective substitutions are established between two subsets of nodes $B_1 \subseteq V_1$ and $B_2 \subseteq V_2$ with equal size $|B_1| = |B_2|$ [95]. Similar to BP, the remaining nodes of $V_1 \backslash B_1$ of graph g_1 are deleted, while the remaining nodes $V_2 \backslash B_2$ of graph g_2 are inserted [95]. Hence, two node assignments ψ_1 and ψ_2 can be computed, i.e. one from g_1 to g_2, and one from g_2 to g_1 [95]. Note that both assignments correspond to valid and complete edit paths between the graphs under consideration.

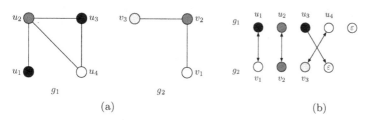

Fig. 5.10: Bijective node assignments (see (b)) of two graphs (see (a)) given by BP2.

The node assignments ψ_1 and ψ_2 are optimised with respect to the HED cost function $f(u,v)$ (see Eq. 5.5) by means of Algorithm 5 [95]. This particular algorithm results in a greedy assignment of local structures rather than an optimal assignment obtained via LSAP solving algorithm as used for BP [95]. That is, for each node $u \in V_1$, the minimum node assignment

---

Algorithm 5 Node assignment ψ

In: Source graph g_1, Target graph g_2, Cost function $\mathcal{C}$
Out: Node Assignment ψ
1: **function** $\psi(g_1, g_2)$
2:      $\psi = \{\}$
3:      **for all** $u \in V_1$ **do**
4:         $v = \varepsilon$
5:         **for all** $w \in V_2$ **do**
6:             **if** $2f(u,w) < f(u,\varepsilon) + f(\varepsilon,w)$ **and** $f(u,w) < f(u,v)$ **then**
7:                $v = w$
8:         $\psi = \psi \cup \{(u \rightarrow v)\}$
9:         $V_1 = V_1 \backslash u$
10:        $V_2 = V_2 \backslash v$
11:      **for all** $v \in V_2$ **do**
12:         $\psi = \psi \cup \{(\varepsilon \rightarrow v)\}$
13:      **return** ψ

$(u \rightarrow v)$ is found with respect to the cost $f(u,v)$ and added to the node assignment ψ (see lines 3 to 8 of Algorithm 5). Note that the substitution $(u \rightarrow v)$ has to satisfy the additional condition [95]

$$2f(u,v) < f(u,\varepsilon) + f(\varepsilon,v)$$

which tests whether the cost for substitution is less than the cost for deletion and insertion of the corresponding nodes. The factor of 2 is needed because, in contrast to HED, the substitution is bijective [95]. If a substitution with a node in V_2 can be established, the corresponding nodes are removed from V_1 and V_2, respectively (see lines 9 and 10 of Algorithm 5) [95]. That is, this assignment is established by means of a suboptimal greedy algorithm [95]. Otherwise, a deletion $(u \rightarrow \varepsilon)$ is added to ψ and V_2 remains unchanged (u is still removed from V_1). Once all nodes in V_1 have been processed, an insertion $(\varepsilon \rightarrow v)$ is added to ψ for all remaining nodes in V_2 (see lines 11 and 12 of Algorithm 5).

Algorithm 5 is run twice, once run with g_1 as source graph and g_2 as target graph and once with g_2 and g_1 as source and target graph, respectively [95]. Eventually, we can use both the node assignments ψ_1 and ψ_2 derived from these runs to infer globally consistent edge edit operations [95]. Hence, similar to BP, we obtain two upper bounds on the true GED [95]. Finally, the distance BP2 is given by the minimum edit distance inferred by Eq. 5.4. That is, BP2 is defined as [95]

$$BP2(g_1, g_2) = \min(d_{\psi_1}, d_{\psi_2}) \quad ,$$

where d_{ψ_1} and d_{ψ_2} refer to costs of globally consistent edit paths ψ_1 and ψ_2 (see Eq. 5.4).

5.4 Polar Graph Dissimilarity (PGD)

In the case of large documents and/or large graphs, suboptimal approaches for GED with cubic or quadratic time complexity can still be a limiting factor for KWS. For this reason, we propose a novel linear time graph matching algorithm in this section. This particular algorithm is applicable to graphs with nodes labelled by (x, y)-coordinates — so called *geometric graphs* — only. The basic idea is that graphs are explicitly embedded in a polar coordinate system, where a dissimilarity method measures the distance between histograms of spatial graph segments. In particular, a graph is *quantised* twofold. First, a graph is segmented into polar segments. Second, each segment is represented by a fixed feature vector derived on the subgraph of the segment. Hence, a complete graph can be represented by concatenating the single feature vector as a fixed-size histogram. Finally, the graph dissimilarity is given by measuring the dissimilarity of their corresponding histograms. We denote this graph dissimilarity by *polar graph dissimilarity* (PGD) from now on.

PGD is inspired by the scale-invariant shape descriptor *contour points distribution histogram* (CPDH) for matching 2D-shape images [297]. Basically, this shape descriptor segments equidistant contour points by means of the polar coordinate system. Thus, a contour image can be formally described by a histogram CPDH $= \{h_1, \ldots, h_i, \ldots, h_n\}$ where h_i consists of the number of contour points in the i-th segment. Finally, two shape images can be compared by computing the *earth mover distance* (or similar metrics) between shifted and mirrored histograms [298].

In the following subsections, we show how this concept can be adapted to graphs $g = (V, E, \mu, \nu)$ with $L_V = \mathbb{R}^2$ [7]. The procedure consists of three subsequent steps. First, graphs are segmented in a polar coordinate system (detailed in Subsection 5.4.1). Second, histograms that represent the node and edge distributions (introduced in Subsection 5.4.2) are extracted for each segment. Finally, two graphs are compared on the basis of the resulting histograms (discussed in Subsection 5.4.3).

---

[7]By means of appropriate hashing (quantisation) functions, arbitrary graphs could be transformed and compared by means of histograms. In particular, such a hashing should allow a graph to be embedded into a fixed-sized histogram while keeping most of the structural representativeness.

5.4.1 Polar Segmentation of Graphs

We transform a given graph g into a polar coordinate system based on its centre of mass (x_m, y_m) as illustrated in Fig. 5.11a[8]. Formally, the (x, y)-coordinates of each node label $\mu(v) = (x, y) \in \mathbb{R}^2$ are transformed to $\mu(v) = (\rho, \theta)$ with

$$\rho = \sqrt{(x - x_m)^2 + (y - y_m)^2} \text{ and } \theta = \text{atan2}((y - y_m)/(x - x_m)) \quad .$$

Clearly ρ denotes the radius from the centre of mass of g to the node's position and $-\pi \leq \theta < \pi$ refers to the angle from the x-axis to the node's position (computed via arctangent function with two arguments in order to return the correct quadrant).

Next, a bounding circle C with radius $\rho_{\max}$ is defined such that C surrounds all nodes of graph g (see Fig. 5.11a for an illustrative example). Based on this bounding circle C, the graph is segmented into $P_r \times P_\phi$ bins where P_r and P_ϕ define the number of different radii and angles, respectively. Note that the centre of mass can be influenced by outliers in the graph. However, this effect is minimised by the polar segmentation, which allows a certain tolerance given by the dimensions of a bin.

An example is given in Fig. 5.11b, where a graph is segmented into 24 bins (with $P_r = 3$ and $P_\phi = 8$). Note that every bin b_i is defined by two radii $\rho_{i_{\min}}$ and $\rho_{i_{\max}}$, and two angles $\theta_{i_{\min}}$ and $\theta_{i_{\max}}$. Hence, every node $v \in V$ with polar coordinates (ρ, θ) and $\rho_{i_{\min}} \leq \rho < \rho_{i_{\max}}$ and $\theta_{i_{\min}} \leq \theta < \theta_{i_{\max}}$ can be assigned to the corresponding bin b_i.

5.4.2 Histogram-Based Representation

Based on the polar segmentation of graphs, graphs can be quantised to fixed-sized histograms. The intrinsic idea behind this procedure is to keep as much of the structural information given by a graph, while embedding the graph to a smaller dimensional histogram. In contrast to existing graph embedding approaches (e.g. [198]), this embedding is directly derived on the segmented graph by means of the node and edge distribution. Hence, the graph embedding is not dependent on measuring graph dissimilarities to a set of prototype graphs by means of a non-linear GED approximation.

Two different histograms can be derived.

---

[8]Node coordinates are *a priori* denormalised by the standard deviation of all node coordinates, for further details we refer to [85].

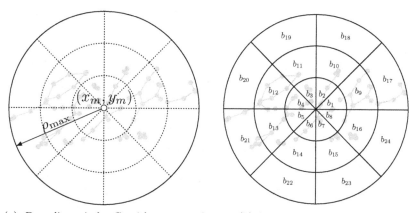

(a) Bounding circle C with centre of mass (x_m, y_m) and radius $\rho_{\max}$

(b) Segmentation of C into bins

Fig. 5.11: Segmentation of graph g by means of a polar coordinate system.

- **Node-Based Histograms**: The first histogram is created by counting the number of nodes per bin. That is, in the histogram $H = \{h_1, \ldots, h_n\}$, h_i represents the node frequency per bin b_i (see Fig. 5.12a for an illustrative example). The resulting histograms are normalised by the $l1$-norm. Formally,

$$\hat{h}_i = \frac{h_i}{\sqrt{\sum_{i=1}^{n} h_i}} \quad,$$

where $\hat{h}_i$ is the normalised bin.

- **Edge-Based Histograms**: The second histogram reflects the distribution of both nodes and edges. To this end, we adapt the concept of *histograms of oriented gradients* (HoG) to (undirected) graphs to form a histogram with radial directions of the corresponding edges. In particular, we first define the maximal number of subbins P that defines the radial range of every subbin b_{n_i}. For each edge in a segment, we measure the Euclidean distance d between the two adjacent nodes as well as the angle θ of the edge to the x-axis. Next, d is assigned to the two enclosing subbins b_{n_i} and b_{n_j} with respect to their radial difference to θ. Formally,

$$b_{n_i} \mathrel{+}= 1 - \frac{\theta - \theta_i}{v} d \quad \text{and} \quad b_{n_j} \mathrel{+}= \frac{\theta - \theta_i}{v} d \quad.$$

Note that every edge is taken into account in both directions as we make use of undirected edges. Finally, the resulting histogram $\{b_{n_1}, \ldots, b_{n_P}\}$ (i.e. one histogram per segment with P bins) is first concatenated to form one global histogram $H = \{h_1 = \{b_{1_1}, \ldots, b_{1_P}\}, \ldots, h_n = \{b_{n_1}, \ldots, b_{n_P}\}\}$ as illustrated in Fig. 5.12b and then normalised by the $l1$-norm.

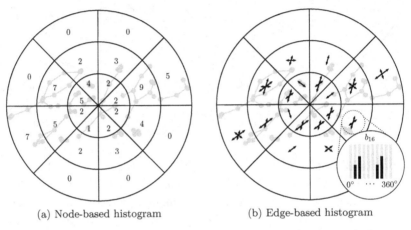

(a) Node-based histogram   (b) Edge-based histogram

Fig. 5.12: Construction of node and edge-based histograms, respectively.

If we compare both types of histograms, we can conclude that node-based histograms lead most probably to a larger structural information loss. That is, the number of nodes per segment can easily be influenced by outliers in the graph. Moreover, the histograms of two different graphs could be very similar just because of the fact that the number of nodes per subgraph and segment is very similar. In contrast to this, edge-based histograms take both the direction and length of edges into account. That is, the likelihood that histograms of two different graphs become similar decreases.

5.4.3 *Histogram-Based Dissimilarity Measure*

To measure the dissimilarity between two histograms H_1 and H_2 that represent the node or the node and edge distribution of two graphs g_1 and g_2, respectively, we make use of the χ^2 distance. However, rather than directly comparing two histograms, we first make use of a *quadtree* segmentation.

That is, we segment the graphs into smaller subgraphs and measure the dissimilarity between smaller subgraphs, as formalised in Algorithm 6. Note that the quadtree segmentation is known to be sensitive to the choice of the centre of mass (similar to the creation of histograms in Subsection 5.4.1). However, the number of outlier nodes in a word graph (i.e. noise, flourish, etc.) is generally relatively small compared to the total number of nodes. As a result, the centre of mass in word graphs is relatively stable as is therefore the corresponding quadtree segmentation.

First, the procedure is initialised by an external call with $l = 1$ (i.e. $PGD(1, g_1, g_2)$). On the basis of two graphs g_1 and g_2 the histograms H_1 and H_2 are created with respect to P_r and P_ϕ (see line 2 of Algorithm 7)[9]. Next, the χ^2-distance between the two histograms is measured (see line 3). If the current recursion level l is equal to the maximal recursion depth r, the distance is returned (see lines 4 and 5). Otherwise, both graphs g_1 and g_2 are segmented into four independent subgraphs. Each of these subgraphs represents the nodes and edges in one of the four quadrants in circle C (see line 6). Finally, for each subgraph pair, the PGD is measured by means of a recursive function call (see line 7). This procedure is repeated until the current recursion level l is equal to the user-defined maximum depth r.

Algorithm 6 Polar graph dissimilarity (PGD)

In: Graphs g_1 and g_2, Number of radii and segments P_r and P_ϕ, Recursion depth r
Out: Polar graph dissimilarity between graph g_1 and g_2
1: **function** PGD(l, g_1, g_2)
2:     Create H_1 based on g_1, P_r, P_ϕ, and H_2 based on g_2, P_r, P_ϕ
3:     Calculate χ^2-distance $d(H_1, H_2)$
4:     **if** l equal r **then**
5:         **return** d
6:     Segment g_1 and g_2 based on quadtree to $g_{1_1}, g_{1_2}, g_{1_3}, g_{1_4}$ and $g_{2_1}, g_{2_2}, g_{2_3}, g_{2_4}$
7:     **return** $d(H_1, H_2) + (\sum_{i=1}^{4} \text{PGD}(l+1, g_{1_i}, g_{2_i}))$

The number of operations grows in a linear manner with the size of the histogram, which depends on the number of radii and segments P_r and P_ϕ, as well as the recursion depth r. Hence, the complexity of the complete procedure is linear.

---

[9]Note that P_r and P_ϕ can be defined separately for each recursion level.

5.5   Summary

In general, the dissimilarity of two graphs is measured by means of a specific graph-matching algorithm. Graphs can either be matched by means of exact or inexact graph-matching approaches. In the case of exact graph matching, correspondences between similar parts or subparts of graphs are found by means of graph and subgraph isomorphism, respectively. Graph isomorphism ensures a bijective mapping between the identical nodes of two graphs (with respect to both labelling and structure). To this end, pairs of nodes are only mapped if the mapping preserves the edge structure in both graphs. A weaker mapping condition has to be fulfilled with subgraph isomorphism that ensures an injective mapping between a graph g_1 and a subgraph of g_2. With respect to graph dissimilarity measures, graph isomorphism and subgraph isomorphism result in a binary dissimilarity measure (i.e. 0 for isomorphic, 1 for non-isomorphic graphs). In order to make exact graph matching more flexible, several refinements have been proposed, such as, for instance, dissimilarity measures based on maximum common subgraphs, or minimum common supergraphs. These measures offer a more refined approach for graph proximity, however, they are still limited in the sense that large parts of the graphs need to be identical in terms of their labels and structures.

Inexact graph matching allows matchings between two graphs that have nothing in common by endowing a certain error-tolerance with respect to both structure and labelling. A powerful and flexible approach for inexact graph matching is provided by *graph edit distance* (GED). GED basically measures the minimum amount of distortion needed to transform one graph into another, given a set of edit operations for nodes and edges. A set of edit operations that transform a given graph g_1 into another graph g_2 is known as the edit path. Every edit operation has an associated cost, and thus, graph edit distance can be formalised as a problem to minimise the sum of total costs among all possible edit paths. However, the number of possible edit paths is exponential with respect to the number of nodes. As a result, exact graph edit distance is intractable for large graphs. In order to solve this problem, several fast but suboptimal algorithms for graph edit distance have been proposed in the last decade.

The *bipartite graph edit distance* (BP) is a prominent suboptimal algorithm for GED with cubic time complexity. BP reformulates the problem of GED to an instance of a *linear sum assignment problem* (LSAP). A minimum cost assignment of local substructures from both graphs can be derived by means of LSAP-solving algorithms. This assignment can then be used to derive a globally consistent graph edit distance.

In the case of the quadratic time algorithm *Hausdorff edit distance* (HED), GED is reformulated as a set matching problem between local substructures. Due to its low computational complexity, HED can be extended to *context-aware Hausdorff edit distance* (CED). With CED, larger local substructures are considered during the optimisation procedure. Finally, *bipartite graph edit distance 2* (BP2) combines the benefits of HED and BP. That is, BP2 provides the upper bound property of BP and the quadratic time complexity of HED.

In this chapter another inexact graph matching algorithm is proposed, namely *polar graph dissimilarity* (PGD). PGD is a linear time algorithm that first embeds graphs into a polar coordinate system. Based on a segmentation carried out in this polar coordinate system, histograms based on the node and edge distribution can be derived, respectively. Finally, a graph dissimilarity can be derived by means of the χ^2 distance measured on the histograms.

Chapter 6

Graph-Based Keyword Spotting

6.1 Overview of our Framework for KWS

Based on the methods and results described in detail in the previous chapters, we are now able to condense the individual parts (i.e. image preprocessing on documents, graph extraction on word images, and graph matching) into a unified graph-based *keyword spotting* (KWS) framework as illustrated in Fig. 6.1.

During the image preprocessing (see (1) in Fig. 6.1) the original document images are processed in order to minimise the influence of variations that are caused, for instance, by noisy backgrounds, skewed scanning, or document degradation. In particular, we enhance edges locally using a *difference of Gaussians* (DoG) in order to address the issue of noisy background and binarise the images by means of a global threshold. Next, single word images are automatically segmented based on projection profiles and, if necessary, manually corrected[1]. Moreover, we correct the skew, i.e. the inclination of the document on single word images, and finally, we skeletonise the results by means of a thinning operator. The technical details and a detailed discussion of the preprocessing steps can be found in Chapter 3.

On the basis of these particular word images, graphs are extracted by means of different graph extraction algorithms (see (2) in Fig. 6.1). To this end, we employ four graph representation formalisms introduced in Chapter 4. These graph representations aim to structure particular characteristics of a word image by means of nodes and edges. In particular, the characteristics of a word image can be seen as the minimal amount of information needed to represent the inherent topological structure of

---

[1]This KWS approach ignores any segmentation errors and can therefore be seen as an upper-bound solution. That is, we focus on KWS systems that generally operate on perfectly segmented word images.

handwriting. One graph representation, for instance, uses nodes in order to represent the location of characteristic points on the skeletonised word image, while edges are used to encode the connectivity of pairs of nodes. Another graph representation is based on a grid-wise segmentation of word images. Nodes are then used to represent the (x, y)-coordinates of the segment's centre of mass, while edges are inserted between two neighbouring segments. Two other graph extractions are then based on adaptive and more flexible segmentations of word images. In order to improve the comparability between graphs of the same word class, the labels $\mu(v)$ of the nodes $v \in V$ are normalised by means of the mean and standard deviation of all (x, y)-coordinates of all nodes in the graph under consideration. Formally, we compute

$$\hat{x} = \frac{x - \mu_x}{\sigma_x} \text{ and } \hat{y} = \frac{y - \mu_y}{\sigma_y} \quad ,$$

where $\hat{x}$ and $\hat{y}$ denote the new node coordinates, while x and y denote the original node position. The value pairs (μ_x, μ_y) and (σ_x, σ_y) represent the means and standard deviations of all (x, y)-coordinates in the graph under consideration.

Finally, the query graphs are matched with the document graphs (see (3) in Fig. 6.1). That is, in our framework for spotting keywords, a query graph q (used to represent a certain keyword) is matched pairwise against all graphs $G = \{g_1, \ldots, g_N\}$ stemming from the underlying document.

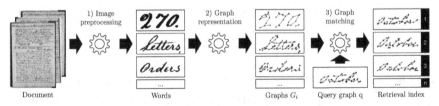

Fig. 6.1: Process of graph-based keyword spotting of the word "October". Note that in contrast to Fig. 2.1, word images are represented by graphs and querying is based on matching a query graph q with all document graphs G_i.

In the remainder of this chapter we introduce several adaptations and extensions of this basic framework that have to be made with respect to the specific task of KWS. In particular, in Section 6.2, we introduce the basic cost model used for graph matching in our KWS framework, show how the raw dissimilarity obtained by graph matching can be normalised, and lastly, we define how retrieval indices can be built on the basis of normalised graph dissimilarities.

Next, in Section 6.3, we introduce three seminal extensions of our basic framework to improve our system with respect to both runtime and accuracy. That is, we first introduce a generic approach for recursively matching parts of graphs rather than complete graphs. This approach can now be applied with any of the graph dissimilarity measures discussed in the previous chapter (*bipartite graph edit distance* (BP), *Hausdorff edit distance* (HED), *context-aware Hausdorff edit distance* (CED), *bipartite graph edit distance 2* (BP2), and *polar graph dissimilarity* (PGD)). The rationale for this recursive partial matching is that it might be beneficial to match several small graphs (and aggregate the matching scores) rather than matching one larger graph. Second, we show how the fast linear time graph matching algorithm of PGD can be used in conjunction with a more accurate graph matching scheme. That is, we show how two matching algorithms can be combined in order to decrease the processing time. Third, we show how the different graph dissimilarities obtained on different graph representations can be combined in order to make the resulting KWS more stable and robust against intraword variations.

Finally, in Section 6.4, the evaluation metrics for KWS (used in the next chapter) are discussed. These metrics are defined for two general scenarios, namely local and global thresholds for KWS.

6.2 From Graph Edit Distance (GED) to Retrieval Indices

This book focuses on the concept of *graph edit distance* (GED), a powerful yet flexible inexact graph matching paradigm. In order to make this specific dissimilarity model applicable in our framework for KWS, we use various fast suboptimal algorithms for the computation of GED.

In the following two subsections, we propose an adaptation of GED with respect to this KWS scenario. First, we introduce the flexible cost model that allows us to handle variations in all of our handwriting graphs (regardless of the actual representation). Second, we build retrieval indices based on the computed and normalised distances.

6.2.1 *Cost Model for Keyword Spotting*

For our cost model we use a weighting parameter $\alpha \in [0, 1]$ that controls whether the edit operation cost on the nodes or on the edges is more important. That is, the cost of any node operation is multiplied by α, while edge operation costs are multiplied by $(1 - \alpha)$. Thus, a setting of $\alpha = 0.5$ leads to balanced importance between node and edge operation cost.

In our framework a constant cost for node deletions and insertions is defined by

$$c(u \rightarrow \varepsilon) = c(\varepsilon \rightarrow v) = \tau_v \in \mathbb{R}^+$$

for any $u \in V_1$ and $v \in V_2$. The same accounts for the edges. That is, we use constant cost $\tau_e \in \mathbb{R}^+$ for edge deletions and insertions. The cost for node substitutions should reflect the dissimilarity of the associated label attributes. In our application the nodes are labelled with (x, y)-coordinates and we use a weighted Euclidean distance on these labels to model the substitution cost. Formally, the cost for a node substitution $(u \rightarrow v)$ with $\mu_1(u) = (x_i, y_i)$ and $\mu_2(v) = (x_j, y_j)$ is defined by

$$c(u \rightarrow v) = \sqrt{\beta(\sigma_x(x_i - x_j))^2 + (1 - \beta)(\sigma_y(y_i - y_j))^2} \quad ,$$

where $\beta \in [0, 1]$ denotes a parameter to weight the importance of the x- and y-coordinate of a node, while σ_x and σ_y denote the standard deviation of all node coordinates in the current query graph. The rationale for this weighting is as follows: The larger the deviation in the x- or y- direction, the more important might be the particular direction and it is thus weighted accordingly. In Table 6.1, the parameters for our cost model are summarised.

Table 6.1: Parameters of the cost model.

Parameter	
α	Weighting factor between node and edge edit costs
β	Weighting factor between x- and y-coordinates
τ_v	Constant costs for node deletions and insertions
τ_e	Constant costs for edge deletions and insertions

In Figs. 6.2, 6.3 and 6.4, we visually investigate the influence of the different cost model parameters by matching two graphs of the word "October". That is, we make use of fixed parameter settings ($\alpha = 0.5$, $\beta = 0.5$, $\tau_v = 6$, and $\tau_e = 6$) while alternating one of the parameters.

First, we investigate the influence of the weighting factor between node and edge edit costs α in Fig. 6.2. We observe that the number of insertions, deletions and substitutions remains almost the same over all three α-levels.

However, there is certainly a difference with respect to node substitutions if we compare the letters 'O' in Fig. 6.2a and 6.2c. If the edge edit operations are more relevant (i.e. $\alpha = 0.1$), substitutions are more often incorrect. In contrast to this, if node edit operations are more relevant (i.e. $\alpha = 0.9$), substitutions become more correct. This is due to the fact that node edit operations are dependent on the (weighted) Euclidean distance, while edge edit operations are dependent on the number of adjacent edges (which is mostly two in this specific graph).

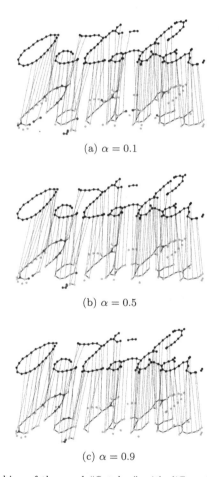

(a) $\alpha = 0.1$

(b) $\alpha = 0.5$

(c) $\alpha = 0.9$

Fig. 6.2: Graph matching of the word "October" with different α-levels. In the lower graph, insertions are marked in dark grey, while deletions and substitutions are marked with light grey nodes. Note that substitutions between source and target graphs are visualised with a line between the involved nodes.

In Fig. 6.3, we observe the influence of the weighting factor β between x- and y-coordinates. Similarly to α, the number of insertions, deletions and substitutions remains almost the same over all three parameter levels. If the x-direction is less relevant and thus the y-direction is more relevant ($\beta = 0.1$), we observe that substitutions in the letter 'O' in Fig. 6.3a are more correct than in Fig. 6.3c. Hence, if the y-direction is more relevant, structures that have only a small vertical difference between two different strokes can also be matched correctly.

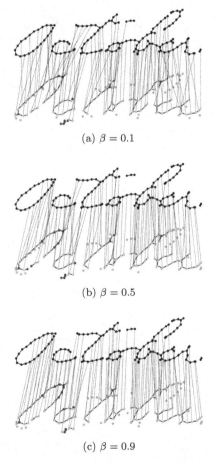

(a) $\beta = 0.1$

(b) $\beta = 0.5$

(c) $\beta = 0.9$

Fig. 6.3: Graph matching of the word "October" with different β-levels. In the lower graph, insertions are marked in dark grey, while deletions and substitutions are marked with light grey nodes. Note that substitutions between source and target graphs are visualised with a line between the involved nodes.

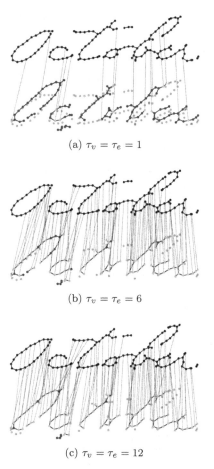

(a) $\tau_v = \tau_e = 1$

(b) $\tau_v = \tau_e = 6$

(c) $\tau_v = \tau_e = 12$

Fig. 6.4: Graph matching of the word "October" with different costs for node/edge deletions and insertions τ_v and τ_e. In the lower graph, insertions are marked in dark grey, while deletions and substitutions are marked with light grey nodes. Note that substitutions between source and target graphs are visualised with a line between the involved nodes.

Finally, the influence of different costs for node and edge deletions and insertions τ_v and τ_e is shown in Fig. 6.4. We observe that the number of substitutions clearly increases if τ_v and τ_e increase, while the number of insertions and deletions decreases, see Fig. 6.4a and 6.4b. That is, node substitution costs that are dependent on the (weighted) Euclidean distance become smaller than the costs for insertions and deletions, and thus, substitutions become more probable. However, we can also observe that certain substructures are wrongly matched with larger τ_v and τ_e (see Fig. 6.4c), in particular, certain nodes in the upper part of the letter 'O' that are substituted with the letter 't'. Hence, we conclude that τ_v and τ_e both directly influence the degree of inexactness, and thus a good compromise between tight matching (i.e. $\tau_v = \tau_e = 1$) and loose matching (i.e. $\tau_v = \tau_e = 12$) is needed (see validations of parameters in Section 7.2.1).

6.2.2 *Retrieval Indices*

To spot keywords, we build two retrieval indices that are optimised separately for a *local* and *global* threshold scenario. Local thresholds are used in the case of a vocabulary of common keywords, while a global threshold is used for arbitrary out-of-vocabulary keywords. In the case of local thresholds, the level of accuracy is measured independently for every query word, while in the case of global thresholds, the accuracy is measured for every query word with the same single threshold[2]. Global thresholds are thus regarded as more realistic, yet also a more difficult scenario.

To build the retrieval indices we first normalise the graph edit distances d_{GED} between query graph q and all document graphs $G = \{g_1, \ldots, g_N\}$ by the sum of the maximum cost edit path between q and g_i, i.e. the sum of the edit path that results from deleting all nodes and edges of q and inserting all nodes and edges in g_i[3]. Formally,

$$d_{\max}(q, g_i) = (|V_q| + |V_{g_i}|)\,\tau_v + (|E_q| + |E_{g_i}|)\,\tau_e \quad .$$

The normalised distance $\hat{d}_{\text{GED}}(q, g_i)$ is then defined as

$$\hat{d}_{\text{GED}}(q, g_i) = \frac{d_{\text{GED}}(q, g_i)}{d_{\max}(q, g_i)} \quad .$$

In cases where a query consists of a set of graphs $\{q_1, \ldots, q_t\}$ that represent the same keyword, the normalised graph edit distance $\hat{d}_{\text{GED}}$ is given by the minimal distance achieved on all t query graphs, i.e. in this case we compute

$$\min_{q_j \in \{q_1, \ldots, q_t\}} \hat{d}_{\text{GED}}(q_j, g_i) \quad .$$

---

[2]In both cases, the threshold defines whether a document word is regarded relevant for a given query word.

[3]Note that d_{GED} can be computed by any of the suboptimal algorithms introduced in Chapter 5, namely BP, HED, CED, and BP2

Based on normalised distances, the retrieval index for local thresholds is derived by

$$r_1(q, g) = -\hat{d}_{\text{GED}}(q, g) \quad,$$

while the retrieval index for global thresholds is derived by

$$r_2(q, g) = -\frac{\hat{d}_{\text{GED}}(q, g)}{\omega} \quad,$$

where ω is a linear scaling factor based on the mean distance of q to its ten nearest neighbours (denoted by $\bar{d}_{\text{GED}}$) and the minimum mean distance of all available query graphs (denoted by $\bar{d}_{\text{GED}_{\min}}$). Formally, ω is given by

$$\omega = 1 + m(\bar{d}_{\text{GED}} - \bar{d}_{\text{GED}_{\min}}) \quad,$$

where m is a user-defined scaling slope. The basic idea of this procedure is to gradually scale $\hat{d}_{\text{GED}}(q, g)$ depending on the mean distance of q to its ten nearest neighbours. This is expected to reduce the intraclass variance between different queries and improve the accuracy for global thresholds.

6.3 Extensions of the Basic Framework

In this section, different extensions for the basic graph-based KWS framework are presented. It turns out that the basic framework (as introduced in Section 6.1) is limited with respect to runtime, especially when BP is used as the basic matching algorithm. That is, the time complexity of the proposed framework is dependent on the number of nodes to be matched, as shown in Table 6.2. In the case of BP, this complexity is cubic, and thus, the limiting factor in the case of large graphs and/or large numbers of graph matchings. Hence, one can either make use of a graph matching algorithm with a lower time complexity (e.g. HED, CED, BP2, and PGD) or try to reduce the computational burden of BP. In the current section we focus on the latter solution. That is, we present two speed-up approaches. The first idea is based on graph segmentations and aggregated graph matchings, while the second idea is built on a fast rejection method that allows us to filter large numbers of irrelevant graphs. Moreover, the basic framework is also limited by certain variations in the different graph representations (discussed in Section 4.3). That is, one graph representation might be better suited to representing certain characteristics of an item of handwriting than another. To reduce the influence of a single graph representation, we introduce a generic framework to combine different distances obtained on different graph representations.

Table 6.2: Time complexity of different graph matching algorithms.

Algorithm	Time complexity
BP	Cubic
HED	Quadratic
CED	Quadratic
BP2	Quadratic
PGD	Linear

6.3.1 *Quadtree Segementations*

To speed up the basic KWS framework, we propose a quadtree segmentation of the individual graphs and a matching and aggregating algorithm. This divide and conquer approach should allow us to substantially speed up the matching procedure, as graph matchings are computed on many small graphs rather than on one large graph. Note that the quadtree segmentation (i.e. the choice of the centre of mass) is known to be sensitive to noise. However, the number of outliers is relatively small, especially due to the employed image preprocessing. Hence, the centre of mass among different word graphs of the same word class is relatively stable, as is therefore also their quadtree segmentation.

The graph segmentation is carried out as follows: First, the bounding box surrounding a graph g is segmented at the centre of mass (x_m, y_m) into four segments as illustrated in Fig. 6.5a. To make this segmentation more robust against variations in the underlying graphs, we overlap each segment depending on a user-defined factor $o \in\ ]0, 1[$. Parameter o defines the overlap of a segment to its neighbouring segments with respect to the width and height of the corresponding segment. That is, for $o = 0.10$, for instance, the overlapping region is 10% of the size of the corresponding segment. For each of the resulting segments one subgraph is created that includes all nodes (and edges) of the corresponding segment. Hence, we obtain four (not necessarily disjoint) subgraphs. These subgraphs are iteratively segmented at their corresponding centre of mass into further subgraphs until the recursion level l is equal to a maximum recursion depth $r > 0$ (defined by the user). This procedure is illustrated in Fig. 6.5.

The actual procedure for computing a dissimilarity between two graphs g and g' using the proposed segmentation is formalised in Algorithm 7 (termed *quadtree graph matching*). Note that this generic procedure can be applied in conjunction with any of the graph matching models introduced in Chapter 5 (e.g. *bipartite graph edit distance* (BP) or *Haus-*

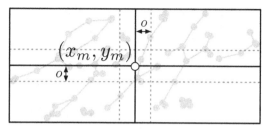

(a) Recursion level $l = 1$ with centre of mass (x_m, y_m) and overlap factor o.

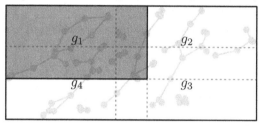

(b) Recursion level $l = 1$ with subgraphs g_1 (highlighted), g_2, g_3, and g_4.

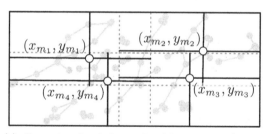

(c) Recursion level $l = 2$ with centres of mass $(x_{m_1}, y_{m_1}), \ldots, (x_{m_4}, y_{m_4})$.

Fig. 6.5: Illustration of quadtree graph segmentation process.

dorff edit distance (HED)). However, in this book we apply this approach in two scenarios only. First, we use a specific version of this algorithm in our linear time graph matching procedure *polar graph dissimilarity* (PGD) (see Section 5.4). Second, we use this segmentation approach with the graph matching algorithm BP, and thus, we denote the resulting method by BP-Q from now on.

The proposed procedure is initialised by an external call with recursion level $l = 1$, i.e. BP-Q$(1, g, g')$. First, both graphs g and g' are segmented into four subgraphs using the procedure described above. Each of these subgraphs represents the nodes and edges in one of the four segments under consideration (with a relative overlap of o) (see line 2 of Algorithm 7). Next, the sum of the four graph edit distances computed on the corresponding four subgraphs are built (see line 3). Finally, the subgraph pairs are further segmented by means of a recursive function call of BP-Q (see line 6). This procedure is repeated until the current recursion level l is equal to the user-defined maximum depth r (see lines 4 and 5).

Algorithm 7 Quadtree graph matching

In: Graphs g and g', Overlap factor o, Maximum recursion depth $r > 0$
Out: Graph distance d_{GED_Q} between graph g and g'
1: **function** BP-Q(l, g, g')
2:    Quadtree segment g and g' to g_1, g_2, g_3, g_4 and g'_1, g'_2, g'_3, g'_4
3:    $d_{\text{GED}_Q} = \sum_{i=1}^{4} d_{\text{GED}(g_i, g'_i)}$
4:    **if** l equal r **then**
5:       **return** d_{GED_Q}
6:    **return** $d_{\text{GED}_Q} + \left(\sum_{i=1}^{4} \text{BP-Q}(l + 1, g_i, g'_i) \right)$

6.3.2 *Fast Rejection Methods*

Our basic KWS framework crucially depends on graph matchings. That is, a document itself as well as the number of queries are represented by a set of graphs $G = \{g_1, \ldots, g_N\}$ and $Q = \{q_1, \ldots, q_n\}$, respectively. In such a scenario, we need to compute $N \times n$ graph matchings. Even with a fast approximation algorithm for the computing of graph dissimilarities, this particular setting might limit the general applicability of our method. One strategy to counteract this problem might be to omit unnecessary computations on graphs that are obviously irrelevant. That is, we aim to reduce the actual number of graph matchings required by efficiently filtering a large number of graphs from G that have a low similarity to the current query graph $q \in Q$. This approach is known as *fast rejection* (FR) [52, 61] and is illustrated as additional step (Filtering) in Fig. 6.6.

We propose to make use of the linear time algorithm *polar graph dissimilarity* (PGD) introduced in Chapter 5 for filtering. We formally proceed as follows: If the distance $d_{PGD}(q, g_i)$ between a query graph q and document graph g_i is below a certain threshold D, we additionally carry out the computationally more expensive (and more accurate) *graph edit distance*

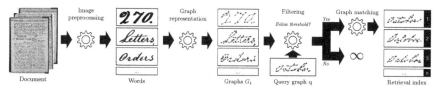

Fig. 6.6: Fast rejection method for graph-based keyword spotting by means of filtering. Note that in contrast to Fig. 2.1, word images are represented by graphs and querying is based on matching a query graph q with all document graphs G_i that are similar enough.

(GED) (denoted by d_{GED}), otherwise we reject graph g_i and assign the graph dissimilarity to be ∞. That is, we define

$$d(q, g_i) \begin{cases} d_{\text{GED}}(q, g_i), & \text{if } d_{\text{PGD}}(q, g_i) < D \\ \infty, & \text{otherwise} \end{cases} \quad . \tag{6.1}$$

Note that one can employ this fast filter processing in conjunction with any graph matching algorithm for GED introduced in Chapter 5. However, in this book we limit this approach to the basic graph matching algorithm *bipartite graph edit distance* (BP), and thus, we denote the resulting method by BP-FRN and BP-FRE when BP is carried out with PGD using node-based and edge-based histograms, respectively.

Clearly, if the threshold D in Eq. 6.1 is increased, the number of filtered document graphs is reduced. Likewise, the number of filtered graphs is increased when threshold D is decreased. Filtering a large number of graphs might have beneficial effects on the runtime of the KWS framework. However, filtering too many, and in particular also relevant graphs at the first step, might crucially deteriorate the KWS accuracy. Hence, the challenge is to find a good tradeoff between high filter rates and low error rates.

6.3.3 Ensemble Methods

Rather than representing word images by only one specific graph representation, one might represent and store both the query graph q and all document graphs $g_i \in G$ with all graph representations as introduced in Chapter 4. In the scheme in Fig. 6.7, a query word is now represented by four graphs q_K, q_G, q_P, and q_S, i.e. one query graph per graph formalism (Keypoint (K), Grid (G), Projection (P), and Split (S)). The same accounts for all document words which are now represented by four sets of document graphs $\{G_K, G_G, G_P, G_S\}$. Hence, rather than matching one query graph against one document graph, one needs to match q_K, q_G, q_P, and q_S with the corresponding sets of document graphs. Consequently, four

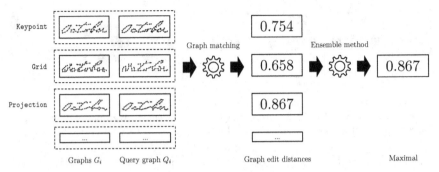

Fig. 6.7: Ensemble method for graph-based keyword spotting by means of the maximal edit distance.

different graph dissimilarities are obtained for each pair of a query word and document word.

Based on this diverse dissimilarity information, we use different combination strategies in order to build a KWS ensemble. Generally, it is known that the combination of multiple classifiers leads to higher accuracy levels than single methods [299]. The motivation in our application is to make the resulting framework more robust with respect to variations of different instances representing the same word.

We propose two strategies for the combination of distances. The first strategy considers all four graph extraction methods by either choosing the minimal, maximal, or mean GED returned on the four representations (termed $d_{\min}$, $d_{\max}$, and d_{mean} from now on). Formally, for one query word q represented by q_K, q_G, q_P, and q_S and one document word g represented by g_K, g_G, g_P, and g_S, we define

$$d_{\min}(q,g) = \min_{i \in \{K,G,P,S\}} d_{\text{GED}}(q_i, g_i) \quad,$$

$$d_{\max}(q,g) = \max_{i \in \{K,G,P,S\}} d_{\text{GED}}(q_i, g_i) \quad,$$

$$d_{\text{mean}}(q,g) = \frac{1}{4} \sum_{i \in \{K,G,P,S\}} d_{\text{GED}}(q_i, g_i) \quad,$$

where d_{GED} is computed by means of *bipartite graph edit distance* (BP), *Hausdorff edit distance* (HED), *context-aware Hausdorff edit distance* (CED), or *bipartite graph edit distance 2* (BP2).

The second strategy considers only two individual graph extraction methods, namely **Keypoint** and **Projection**. In a preliminary evaluation it clearly turns out that these two formalisms offer the greatest potential with respect to distance-based classification (see Section 4.4).

Two different weighted sums are applied to combine the respective distances with each other (termed d_{sum_γ} and $d_{\text{sum}_{\text{map}}}$ from now on). Formally,

$$d_{\text{sum}_\gamma}(q, g) = \gamma \, d_{\text{GED}}(q_K, g_K) + (1 - \gamma) \, d_{\text{GED}}(q_P, g_P) \quad ,$$

$$d_{\text{sum}_{\text{map}}}(q, g) = \delta \, d_{\text{GED}}(q_K, g_K) + \zeta \, d_{\text{GED}}(q_P, g_P) \quad ,$$

where $\gamma \in \,]0, 1[$ denotes a user-defined weighting factor. The parameters δ and ζ for the second sum denote weighting factors that refer to the mean average precision of the individual KWS systems operating on Keypoint and Projection graphs, respectively. Hence, the better a system performs with a specific graph representation, the larger is its weighting factor.

In our proposed system we combine distances obtained from one matching algorithm only. Note, however, that one could also employ different graph-matching algorithms for GED at a time, and thus, combine different graph edit distances obtained by different graph-matching algorithms (on different representations). In this book we focus on ensemble methods that rely on one graph-matching algorithm that is applied to several formalisms.

6.4 Evaluation Metric

In order to measure and compare the KWS accuracy of different systems we compute the number of *true positives* (TP), *false positives* (FP), and *false negatives* (FN) based on an inspected (and if necessary corrected) ground truth[4]

TP = Number of relevant and retrieved words,

FP = Number of irrelevant and retrieved words,

FN = Number of relevant and not retrieved words.

These numbers can then be used to derive *recall* (R) and *precision* (P)

$$R = \frac{TP}{TP + FN} \text{ and } P = \frac{TP}{TP + FP} \quad .$$

Both recall and precision can be computed for two types of thresholds, namely *local* and *global* thresholds. In the case of local thresholds, the KWS performance is measured individually for each keyword and then averaged over all keyword queries. In the case of global thresholds, the same threshold is used for all keywords. This scenario is more practical for a

---

[4]Available at http://www.histograph.ch/

real-world KWS system but requires individual keyword scores to be comparable with each other. Hence, the global threshold scenario is generally more challenging.

Two metrics are used to evaluate the quality of the KWS system. For global thresholds, the *average precision* (*AP*) is measured, which is the area under the recall-precision curve for all keywords given a single (global) threshold. For local thresholds, we compute the *mean average precision* (*MAP*), that is, the mean over the *AP* of each individual keyword query. Both values can be computed, for instance, by using the `trec_eval`[5] software.

In Fig. 6.8, two exemplary query words and the top ten retrieval results are shown. Note that the local threshold is denoted by a dashed line, while the global threshold is denoted by a dotted line.

(a) Query Q_1: *Instructions.*

(b) Query Q_2: *ordered*

Fig. 6.8: Exemplary local (dashed line) and global (dotted line) threshold. True positives are underlined in black and false positives are underlined in grey.

In the case of local thresholds, we observe on query Q_1 and Q_2 an *AP* of

$$AP(Q_1) = \frac{1}{5}\left(\frac{1}{1+0} + \frac{2}{2+0} + \frac{2}{2+1} + \frac{2}{2+2} + \frac{3}{3+2}\right) \qquad = 0.753$$

$$AP(Q_2) = \frac{1}{3}\left(\frac{1}{1+0} + \frac{1}{1+1} + \frac{2}{2+1}\right) \qquad = 0.722$$

---

[5] http://trec.nist.gov/trec_eval

that leads to an MAP over both queries of

$$MAP = \frac{1}{2}\left(AP(Q_1) + AP(Q_2)\right) = \frac{1}{2}\left(0.753 + 0.722\right) = 0.738 \quad .$$

In the case of global thresholds, we observe an AP over both queries of

$$AP = \frac{1}{7}\left(\frac{2}{2+0} + \frac{3}{3+1} + \frac{4}{4+2} + \frac{4}{4+4} + \frac{5}{5+5} + \frac{5}{5+7} + \frac{5}{5+9}\right) = 0.599 \quad .$$

6.5 Summary

The proposed graph-based *keyword spotting* (KWS) framework is based on three basic process steps, namely image preprocessing, graph representation, and graph matching. First, document images are preprocessed and segmented into single word images. Next, handwritten word images are represented by means of a specific graph representation. Third, a query graph is matched pairwise with all documents graphs for spotting keywords.

This chapter first summarises the three steps on a high level. Then, several concepts, adaptations and extensions which are useful for the graph-based KWS framework are introduced and discussed. First, the underlying cost model for KWS based on graph edit distances is introduced and discussed. Next, we show how the resulting graph edit distances can be transformed into a retrieval index for both local and global thresholds. Basically, local thresholds are used in the case of a vocabulary of common keywords, while global thresholds are used for out-of-vocabulary keywords.

Moreover, the basic KWS framework is extended with respect to improvements of both runtime and accuracy. That is, we propose two different speed-up approaches. The first approach makes use of a quadtree graph segmentation. In particular, graphs are segmented into smaller subgraphs, and thus, the graph-matching process is divided and carried out on smaller subgraphs rather than on the complete graph. The second speed-up approach makes use of a novel linear time graph-matching algorithm to filter large numbers of irrelevant graphs. That is, the computationally more expensive *graph edit distance* (GED) is computed if, and only if, the first graph dissimilarity measure is below a certain threshold. This procedure can be seen as a combination of different graph dissimilarity concepts with the main goal of speeding up the KWS process. We also show how different distances, obtained by one graph-matching concept applied to different graph representations, can be combined to build a KWS ensemble. This combinatorial approach could potentially increase the stability and accuracy of the KWS framework.

Last but not least, we discuss our basic quality metric used in the evaluation. The accuracy of a KWS system is typically measured by two different measures, namely *mean average precision* (MAP) in the case of local thresholds and *average precision* (AP) in the case of global thresholds.

Chapter 7

Experiments

7.1 Overview of Experimental Evaluation

For the the purpose of experimental evaluation we consider two well-known manuscripts, namely *George Washington* (GW) and *Parzival* (PAR), as well as two manuscripts from a recent *keyword spotting* (KWS) benchmark competition, namely *Alvermann Konzilsprotokolle* (AK) and *Botany* (BOT) (see Chapter 3 for further details on these documents).

In the next section (Section 7.2), the experimental setup is explained. We start our experimental evaluation by testing the proposed graph- and template-based KWS framework in conjunction with the *bipartite graph edit distance* (BP) algorithm in Section 7.3. Even though this suboptimal algorithm for *graph edit distance* (GED) offers cubic time complexity with respect to the number of nodes of the involved graphs, the computational time might still be a limiting factor in the case of large documents.

Thus, we propose to speed up the whole KWS system by means of alternative matching procedures. In particular, with a first group of speed-up approaches we aim to reduce the computational complexity of BP itself by different heuristics (by using segmentations of the graphs or applying fast filtering of irrelevant graphs by means of the polar graph distance *polar graph dissimilarity* (PGD)). The evaluation of these procedures can be found in Section 7.4.

The second group of speed-up approaches for our KWS system is based on suboptimal algorithms for GED with a lower complexity than BP such as, for example, *Hausdorff edit distance* (HED) or *bipartite graph edit distance 2* (BP2) with quadratic — rather than cubic — time complexity. The lower computational complexity of HED allows us to consider the larger context of the node structure during the matching procedure (termed *context-aware Hausdorff edit distance* (CED)). The empirical results of KWS with these procedures (HED, CED, and BP2) can be found in Sections 7.5 and 7.6. Next, the different graph representations and algorithms for GED are combined to build a KWS ensemble, which is evaluated in Section 7.7.

Moreover, the proposed graph-based KWS system is evaluated in a cross-evaluation scenario. That is, optimal parameter settings from one document are used to evaluate all other documents in order to proof the generalisability of this framework. These results are presented in Section 7.8.

Next, the results of all graph-based KWS systems are summarised and quantitatively and qualitatively compared with each other in Section 7.9. Finally, in Section 7.10 we compare our system with both template-based and learning-based reference systems.

In Table 7.1 an overview of the experimental evaluation is provided, including the employed graph matching algorithms and possible extensions.

Table 7.1: Structure of the experimental evaluation under consideration of the employed graph-matching algorithms and their algorithmic extensions.

Method	Description	Evaluation
Baseline BP	Baseline KWS system using the cubic time algorithm BP (see Section 5.3.2)	Section 7.3
Speed-up BP	Speed-up approaches of baseline KWS system using quadtree graph segmentations (see Section 6.3.1) and fast rejection approaches (see Section 6.3.2)	Section 7.4
Speed-up HED	Speed-up approaches using quadratic time algorithms HED and CED (see Section 5.3.3)	Section 7.5
Speed-up BP2	Speed-up approach using quadratic time algorithm BP2 (see Section 5.3.4)	Section 7.6
Ensemble	Ensemble methods based on different graph representations using BP graph matching and others (see Section 6.3.3)	Section 7.7
Cross-evaluation	Cross-evaluation of graph-based KWS system	Section 7.8
Summary	Quantitative and qualitative summary of all graph-based KWS systems	Section 7.9
Comparison	Comparison of graph-based KWS systems with template-based and learning-based KWS reference systems	Section 7.10

7.2 Experimental Setup

The experimental evaluation is carried out in two steps. First, all parameters are optimised on ten manually selected keywords (with different word lengths), as shown in Fig. 7.1. We define a validation set that consists of all or at least 10 random instances per selected keyword and a maximum of 900 additional random words (in total 1,000 words). Second, the optimised systems are evaluated on the same training and test sets as used in [32] and [80]. All templates of a keyword present in the training set are used for KWS. In Table 7.2, the number of keywords as well as the sizes of the training and test sets for all datasets can be found[1].

Note that the chosen keywords are based on standard benchmark datasets, and thus, they do not necessarily represent their potential users (e.g. theologian, botanist, etc.) very well. However, for reasons of comparability to other systems we follow this practice in our evaluation.

(a) George Washington (GW)

(b) Parzival (PAR)

(c) Alvermann Konzilsprotokolle (AK)

(d) Botany (BOT)

Fig. 7.1: Instances of the selected keywords from the four datasets used for optimisation.

---

[1]GW is divided into four folds, and thus both the both the size of the dataset and the corresponding KWS accuracies are averaged over the four folds.

Table 7.2: Number of keywords and number of word images in the training and test sets
of the four datasets.

Dataset	Keywords	Train	Test
GW	105	2,447	1,224
PAR	1,217	11,468	6,869
AK	200	1,849	3,734
BOT	150	1,684	3,380

7.2.1 *Validation of Parameters*

For every manuscript, all parameters of the graph-based KWS framework
are optimised in six subsequent steps. First, the parameters of the four
different graph representations (introduced in Section 4.3) are optimised.
Next, the cost function for the graph matching algorithms (introduced in
Section 5.3) is optimised using the different graph representations. In three
additional steps, the quadtree segmentation (introduced in Section 6.3.1),
the fast rejection method (introduced in Section 6.3.2), as well as the ensemble methods (introduced in Section 6.3.3) are optimised independently.
Finally, all of the graph-based KWS systems are optimised with respect to
global thresholds (see Section 6.4).

In the following, we summarise the main results of the validation of
the parameters; for details and complete validation results we refer to
Chapter 10.

- **Validation of the Graph Representations**: The parameters for
  each graph representation formalism (see Table 7.3) are optimised
  with respect to the *mean average precision* (MAP) on the validation set using different node and edge deletion/insertion costs $\tau_v =
  \tau_e = \{1, 4, 8, 16, 32\}$ and fixed weighting parameters $\alpha = \beta = 0.5$. In
  Table 7.4, an overview of the tested parameters is provided for all
  four manuscripts. The best performing parameters are marked with an
  asterisk.

  Note that in many cases, parameter settings turn out to be optimal in
  cases of large graphs (i.e. minimal values of D, w, h, D_v, D_h, and D_w).
  That is, larger graphs generally lead to higher accuracy rates. Yet,
  for a good tradeoff between runtime and accuracy, we have to limit
  the lower bound of the range of parameter settings, especially in the
  case of the cubic time graph-matching algorithm BP. Note, moreover,
  that we evaluate all four graph representations of the GW and PAR

manuscripts, while in the AK and BOT manuscripts, only the two best performing graph representations (i.e. `Keypoint` and `Projection`) are evaluated (see preliminary evaluation in Section 4.4).

Table 7.3: Parameters of the four graph representation formalisms.

Method	Parameter	
Keypoint	D = Distance threshold	
Grid	w = Segment width	h = Segment height
Projection	D_v = Vertical threshold	D_h = Horizontal threshold
Split	D_w = Width threshold	D_h = Height threshold

Table 7.4: Optimisation of the graph representation formalisms. Optimal parameters are marked with an asterisk.

Method		Parameter
GW	Keypoint	$D = \{4^*, 6, 8, 10, 12\}$
	Grid	$w = \{6^*, 8, 10, 12, 14\} \times h = \{6^*, 8, 10, 12\}$
	Projection	$D_v = \{4^*, 6, 8, 10\} \times D_h = \{4, 6^*, 8, 10\}$
	Split	$D_w = \{4, 6^*, 8, 10\} \times D_h = \{4^*, 6, 8, 10\}$
PAR	Keypoint	$D = \{2^*, 4, 6, 8, 10, 12\}$
	Grid	$w = \{4^*, 6, 8, 10, 12\} \times h = \{4^*, 6, 8, 10\}$
	Projection	$D_v = \{2^*, 4, 6, 8, 10\} \times D_h = \{4^*, 6, 8, 10\}$
	Split	$D_w = \{2^*, 4, 6, 8, 10\} \times D_h = \{4^*, 6, 8, 10\}$
AK	Keypoint	$D = \{16^*, 20, 24, 28, 32, 36\}$
	Projection	$D_v = \{10, 12, 14, 16^*\} \times D_h = \{10^*, 12, 14, 16\}$
BOT	Keypoint	$D = \{16^*, 20, 24, 28, 32, 36\}$
	Projection	$D_v = \{10, 12, 14^*, 16\} \times D_h = \{10, 12^*, 14, 16\}$

- **Validation of the Cost Function for Graph Edit Distance**: Using the optimal parameters for each graph representation formalism, the cost functions for graph edit distance are optimised. That is, we evaluate 25 pairs of constants for node and edge deletion/insertion costs ($\tau_v = \tau_e = \{1, 4, 8, 16, 32\}$) in combination with all weighting parameters $\alpha = \{0.1, 0.3, 0.5, 0.7, 0.9\}$ and $\beta = \{0.1, 0.3, 0.5, 0.7, 0.9\}$. Hence, we evaluate a total of $5 \times 5 \times 5 \times 5 = 625$ parametrisations per graph representation and manuscript (resulting in 5,000 settings in total). This optimisation is individually repeated for the graph-matching algorithms BP, HED, CED, and BP2. Note that we use the same cost function for BP-Q, BP-FRN, and BP-FRE as used for BP. In Table 7.5, the optimal cost function parameters are shown for the BP graph-matching algorithm and the corresponding graph representations. For all other graph-matching algorithms we refer to Table 10.1 in the Chapter.

Table 7.5: Optimal cost function parameter for graph edit distance computation.

Method	GW				PAR				AK				BOT			
	τ_v	τ_e	α	β	τ_v	τ_e	α	β	τ_v	τ_e	α	β	τ_v	τ_e	α	β
							BP									
Keypoint	4	1	0.5	0.1	4	4	0.5	0.3	16	16	0.5	0.1	32	32	0.3	0.1
Grid	4	1	0.7	0.1	4	1	0.7	0.5	-	-	-	-	-	-	-	-
Projection	4	1	0.5	0.1	4	1	0.5	0.5	8	32	0.7	0.1	8	32	0.9	0.3
Split	4	1	0.5	0.1	4	1	0.3	0.3	-	-	-	-	-	-	-	-

- **Validation of the Quadtree Segmentation Method**: For the validation of the quadtree segmentation approach BP-Q, two parameters are optimised, namely the maximum recursion depth r and the overlap factor o. We evaluate five maximum recursion depths $r \in \{1, 2, 3, 4, 5\}$ in combination with 20 overlap factors $o \in \{0.01, 0.02, \ldots, 0.20\}$. In Table 7.6 the best performing parameters are presented for Keypoint and Projection graphs stemming from different manuscripts (for this particular evaluation only two out of four graph representations are tested).

Table 7.6: Optimal recursion depth r and overlap factor o for quadtree segmentations.

Method	GW		PAR		AK		BOT	
	r	o	r	o	r	o	r	o
Keypoint	1	0.01	1	0.01	5	0.00	1	0.14
Projection	1	0.02	1	0.01	2	0.03	3	0.08

- **Validation of the Fast Rejection Method**: The validation of the fast rejection methods BP-FRN and BP-FRE is a two-step procedure. First, parameters of PGD are optimised with respect to *average pre-*

cision (AP). That is, different polar segmentations (defined via P_r and P_ϕ) are validated for two recursion levels l (i.e. we define the maximal recursion depth to $r = 2$). For $l = 1$, the parameter combinations $P_r = \{1, 2, 3, 4, 5, 6\} \times P_\phi = \{4, 8, 12, 16, 20, 24, 28, 32, 36, 40\}$ are evaluated, while for $l = 2$ the parameter combinations $P_r = \{1, 2, 3, 4\} \times P_\phi = \{2, 4, 6, 8, 10\}$ are evaluated. Hence, in total, we evaluate $6 \times 10 \times 4 \times 5 = 1{,}200$ parameter combinations for every graph representation. Note that for PGD-Edge we use 10 radial bins in all our experiments. In Table 7.7 the best performing parameters are presented for all graph representations stemming from different manuscripts.

Table 7.7: Optimal P_r and P_ϕ for PGD on both recursion levels l in conjunction with node- and edge-based histograms, respectively.

Method		PGD-Node				PGD-Edge			
		$l = 1$		$l = 2$		$l = 1$		$l = 2$	
		P_r	P_ϕ	P_r	P_ϕ	P_r	P_ϕ	P_r	P_ϕ
GW	Keypoint	5	8	1	4	4	16	1	4
	Grid	4	8	1	2	5	40	1	4
	Projection	5	16	1	4	6	4	1	2
	Split	6	24	1	2	5	40	1	4
PAR	Keypoint	6	36	4	8	3	36	3	4
	Grid	1	36	1	8	6	36	4	4
	Projection	6	36	4	4	4	36	4	8
	Split	2	36	3	8	6	8	3	8
AK	Keypoint	4	20	1	2	4	4	1	4
	Projection	4	20	1	10	4	20	2	4
BOT	Keypoint	6	40	1	4	1	16	2	4
	Projection	4	40	4	4	1	36	2	4

In a second step, we optimise the threshold D that controls the number of filtered graphs. For BP-FRN we evaluate thresholds $D = \{0.75, 1.5, \ldots, 29.25, 30\}$ and for BP-FRE we evaluate thresholds $D = \{2.5, 5, \ldots, 147.5, 150\}$. In Figs. 7.2 and 7.3, the AP as well as the *filter rate* (FR) are shown as a function of rejection threshold D on the GW manuscript for BP-FRN and BP-FRE, respectively. By increasing D we observe that the KWS performance is generally improved. Simultaneously, the number of filtered graphs decreases (generally making the

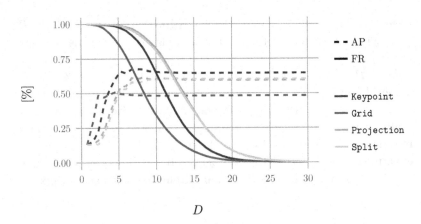

Fig. 7.2: GW: AP and FR for BP-FRN as a function of threshold D.

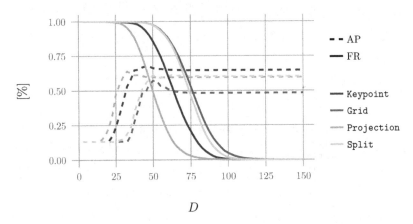

Fig. 7.3: GW: AP and FR for BP-FRE as a function of threshold D.

KWS process slower). Threshold D is determined such that the AP is maximised (if this threshold is too restrictive, we choose the next highest threshold where the AP does not further decrease). Similar effects can be observed on the other manuscripts (see Chapter 10 for details). In Table 7.8, the rejection threshold D and the corresponding FRs are shown for each graph representation and manuscript.

- **Validation of the Ensemble Methods**: For the KWS ensemble methods the weighting factor $\gamma \in \{0.05, \ldots, 0.95\}$ for the ensemble sum$_\gamma$ is the sole parameter that needs to be optimised (all other en-

Table 7.8: Optimal D and corresponding FR.

Method		BP-FRN D	BP-FRN FR	BP-FRE D	BP-FRE FR
GW	Keypoint	7.50	88.08	45.0	93.57
	Grid	4.50	89.43	55.0	94.17
	Projection	8.25	91.55	35.0	92.66
	Split	12.75	56.74	62.5	82.75
PAR	Keypoint	24.00	82.49	65.0	94.12
	Grid	6.75	91.91	77.5	94.48
	Projection	17.25	95.51	82.5	91.00
	Split	16.50	88.42	65.0	94.66
AK	Keypoint	7.50	96.00	60.0	89.18
	Projection	12.75	92.36	110.0	90.03
BOT	Keypoint	13.50	95.97	92.5	84.59
	Projection	19.50	91.09	105.0	93.78

semble strategies need no additional parameter tuning). In Table 7.9, the optimal γ is given for all graph matching algorithms on all four manuscripts.

Table 7.9: Optimal γ for the sum$_\gamma$ ensemble method.

Method	GW	PAR	AK	BOT
BP	0.30	0.10	0.25	0.45
HED	0.10	0.10	0.20	0.75
CED $(n = 2)$	0.25	0.15	0.05	0.90
CED $(n = 4)$	0.25	0.05	0.05	0.05
CED $(n = 6)$	0.30	0.15	0.25	0.05
BP2	0.20	0.50	0.80	0.95

- **Validation of the Global Threshold**: For KWS systems using global thresholds we employ retrieval index r_2 rather than r_1 (see Subsection 6.2.2 for details). The scaling slope m for r_2 is individually optimised for each system. We tested $m \in \{0.005, \ldots, 5.000\}$ for PGD-Node and PGD-Edge, and $m \in \{0.05, \ldots, 50.00\}$ for all remaining graph-matching algorithms. In Tables 7.10 and 7.11, the optimal m for r_2 is given for single and ensemble-based systems using BP graph-matching algorithms, respectively. For validation results on all other single and ensemble-based systems we refer to Tables 10.2 and 10.3, respectively. Note that we use the same values for m for BP-FRN and BP-FRE as used for BP.

Table 7.10: Optimal m for retrieval index r_2 using single graph-matching algorithms.

Method		GW	PAR	AK	BOT
BP	Keypoint	4.55	2.50	3.35	3.30
	Grid	4.70	3.60	-	-
	Projection	4.90	3.90	2.35	6.05
	Split	4.75	2.65	-	-

Table 7.11: Optimal m for retrieval index r_2 using ensemble methods.

Method		GW	PAR	AK	BOT
BP	min	4.25	7.70	2.35	4.15
	max	3.80	5.35	2.00	1.85
	mean	1.90	2.10	1.10	1.75
	sum_γ	4.15	7.15	1.90	5.05
	sum_{map}	2.05	2.30	1.95	2.20

7.3   The Baseline KWS System Using BP

In this first experimental evaluation, we compare different graph representations with each other by means of the baseline KWS system using *bipartite graph edit distance* (BP) on the benchmark manuscripts. We evaluate the graph representations in a KWS experiment with respect to the MAP for local thresholds and the AP for global thresholds, as shown in Table 7.12[2]. For every graph representation and manuscript available, the mean number of nodes $|\bar{V}|$ and edges $|\bar{E}|$ is shown in Table 7.13.

On the GW manuscript, the graph representation Keypoint achieves the highest accuracies in both threshold scenarios. The remaining graph representations clearly result in lower accuracy rates. In particular, in the more difficult global scenario Keypoint outperforms the other approaches. This is particularly interesting as the Keypoint graphs refer to the smallest

---

[2]For better readability, MAP and AP are always represented as percentage (i.e. without explicit %) with an absolute reference of 100%.

Table 7.12: MAP and AP of the KWS system using different graph representations and BP on the GW, PAR, AK, and BOT manuscripts. The best result per column is shown in bold.

Method	GW		PAR		AK		BOT	
	MAP	AP	MAP	AP	MAP	AP	MAP	AP
Keypoint	**66.08**	**54.37**	62.04	61.36	**77.24**	**76.32**	45.06	33.84
Grid	60.02	45.52	56.50	43.55	-	-	-	-
Projection	61.43	48.45	**66.23**	**64.23**	76.02	74.38	**49.57**	**38.33**
Split	60.23	48.41	59.44	58.29	-	-	-	-

Table 7.13: Mean number of nodes $|\bar{V}|$ and edges $|\bar{E}|$ of different graph representations in the GW, PAR, AK, and BOT manuscripts.

Method	GW		PAR		AK		BOT	
	$|\bar{V}|$	$|\bar{E}|$	$|\bar{V}|$	$|\bar{E}|$	$|\bar{V}|$	$|\bar{E}|$	$|\bar{V}|$	$|\bar{E}|$
Keypoint	74	69	81	79	91	91	95	93
Grid	90	88	95	94	-	-	-	-
Projection	74	70	88	90	64	68	64	65
Split	80	77	78	80	-	-	-	-

graphs in our portfolio. In contrast to this, the largest graphs obtained by Grid achieve the lowest accuracies (i.e. about 6 to 9 basis points lower than Keypoint). These results are also reflected in Fig. 7.4, where the recall-precision curves are plotted for both threshold scenarios and all graph representations.

Slightly different observations can be made on the PAR manuscript, where the graph representation Projection leads to the highest accuracy rates (also shown in Fig. 7.5). However, the graph representation Keypoint can achieve comparable rates, while a certain drop in accuracy can be observed on the Grid and Split graphs. Overall, we can observe more distinguishable differences among the different graph representations when compared with GW. In particular, we observe that the Grid graph representation is not able to keep up with the other representations with respect to AP.

On the AK and BOT manuscripts, we only consider the two graph representations performing the best on GW and PAR, namely Keypoint

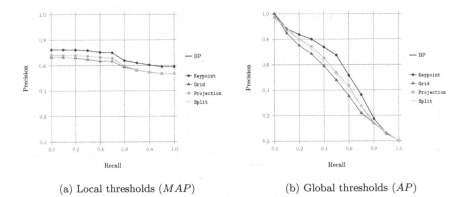

(a) Local thresholds (MAP) (b) Global thresholds (AP)

Fig. 7.4: GW: Recall-precision curves using BP graph-matching algorithm.

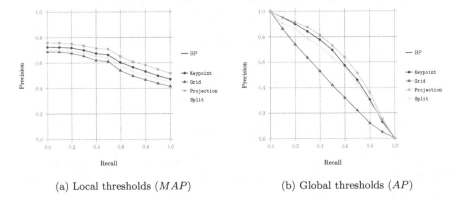

(a) Local thresholds (MAP) (b) Global thresholds (AP)

Fig. 7.5: PAR: Recall-precision curves using BP graph-matching algorithm.

and Projection[3]. We observe slightly higher accuracy levels in the case of Keypoint graphs on AK, while Projection leads to better rates on the BOT manuscript. Note that the method Projection leads to smaller graphs than Keypoint, and thus, also lower matching times can be expected on both manuscripts (the matching times are discussed in more detail in the next section). In Figs. 7.6 and 7.7, the recall-precision curves are shown for AK and BOT, respectively. Overall, we observe clearly higher accuracy rates on AK when compared with BOT. In particular, in the global thresh-

---

[3]For all remaining experiments, all graph representations are employed for GW and PAR, while only Keypoint and Projection are employed for AK and BOT.

old scenario on BOT both graph representations achieve APs of 33% and 38% only.

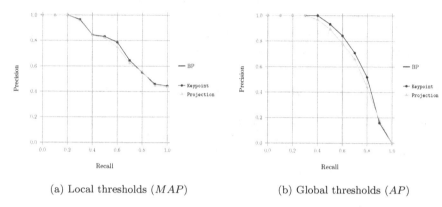

(a) Local thresholds (MAP) (b) Global thresholds (AP)

Fig. 7.6: AK: Recall-precision curves using BP graph-matching algorithm.

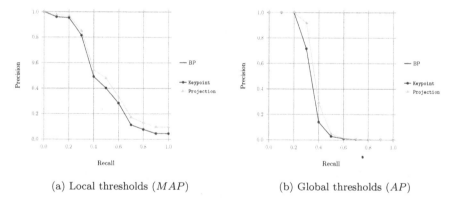

(a) Local thresholds (MAP) (b) Global thresholds (AP)

Fig. 7.7: BOT: Recall-precision curves using BP graph-matching algorithm.

7.4   Speeding Up the Baseline KWS System

In the following two subsections, two different approaches are evaluated for speeding up the baseline KWS system described above. The first speed-up approach is based on a quadtree graph segmentation. That is, the

BP graph matchings are conducted on smaller subgraphs rather than on complete graphs. This divide and conquer approach should allow us to substantially speed up the matching procedure. The second speed-up approach is based on a *fast rejection* (FR) approach. That is, pairs of graphs are first examined by a linear graph dissimilarity. If, and only if, the graphs are similar enough with respect to this dissimilarity, is the computationally more expensive BP graph matching actually applied.

7.4.1   *KWS System Based on BP with Quadtree Segmentations*

First, we examine the speed-up approach BP-Q that makes use of a quadtree segmentation of the graphs under consideration (introduced in Section 6.3.1). Note that for this evaluation we consider Keypoint and Projection graphs only.

Table 7.14: GW: *MAP*, *AP*, and *SF* of the KWS system using different graph representations and BP in conjunction with quadtree segmentations. With ± we indicate the relative percental gain or loss in the accuracy of BP-Q when compared with BP. The best system is shown in bold.

Method	MAP	$\pm$	AP	$\pm$	SF
		BP			
Keypoint	**66.08**		54.37		
Projection	61.43		48.45		
		BP-Q			
Keypoint	65.92	−0.16	**54.43**	+0.06	17
Projection	59.57	−1.86	47.65	−0.80	15

On the GW manuscript, the proposed method BP-Q achieves *speed-up factors* (*SF*s) of about 15 to 17, as shown in Table 7.14[4]. This refers to a substantial reduction in the runtime. In general, a loss of accuracy of BP-Q has to be observed when compared with BP. However, this rather small deterioration seems to be acceptable in view of the substantial gain in runtime. Note, moreover, that Keypoint leads to slightly better results with respect to *AP* using BP-Q rather than BP. The same observations can be made in the recall-precision curves for BP and BP-Q in Fig. 7.8.

---

[4]Note that all performance-related experiments have been measured on the same machine (iMac 5K, 4GHz Intel Core i7, 32GB DDR3) in a single thread scenario.

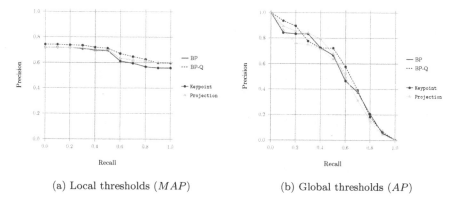

(a) Local thresholds (MAP) (b) Global thresholds (AP)

Fig. 7.8: GW: Recall-precision curves using BP and BP-Q graph-matching algorithms.

Quite similar observations can be made on PAR, as shown in Table 7.15 and Fig. 7.9. For both graph representations speed-up factors of 21 to 22 can be observed. On `Keypoint` graphs we observe a decrease in the MAP and the AP of 4 to 5 basis points. For `Projection` graphs we observe a rather small reduction in the MAP and an increase in the AP when compared with the baseline.

Table 7.15: PAR: MAP, AP, and SF of the KWS system using different graph representations and BP in conjunction with quadtree segmentations. With $\pm$ we indicate the relative percental gain or loss in the accuracy of BP-Q when compared with BP. The best result per column is shown in bold.

Method	MAP	$\pm$	AP	$\pm$	SF
		BP			
Keypoint	62.04		61.36		
Projection	**66.23**		64.23		
		BP-Q			
Keypoint	56.83	-5.21	56.95	-4.41	21
Projection	64.62	-1.61	**67.82**	$+3.59$	22

In contrast to GW and PAR, we observe lower speed-up factors of 10 to 16 on the AK manuscripts as indicated in Table 7.16. A possible reason for this might be larger optimal recursion depths r used on this dataset. Con-

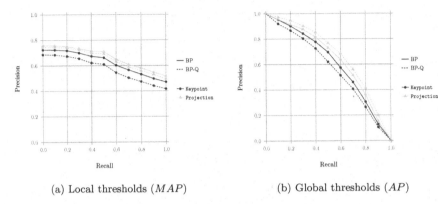

(a) Local thresholds (*MAP*)   (b) Global thresholds (*AP*)

Fig. 7.9: PAR: Recall-precision curves using BP and BP-Q graph-matching algorithms.

Table 7.16: AK: *MAP*, *AP*, and *SF* of the KWS system using different graph representations and BP in conjunction with quadtree segmentations. With ± we indicate the relative percental gain or loss in the accuracy of BP-Q when compared with BP. The best result per column is shown in bold.

Method	*MAP*	±	*AP*	±	*SF*
		BP			
Keypoint	**77.24**		**76.32**		
Projection	76.02		74.38		
		BP-Q			
Keypoint	75.03	−2.21	72.40	−3.92	16
Projection	73.31	−2.71	69.65	−4.73	10

sequently, the number of matchings and therefore also the overall matching time becomes somewhat larger when compared to the manuscripts GW and PAR where a recursion depth of $r = 1$ turns out to be optimal. The reduction in the runtime is accompanied by an accuracy drop of 2 to 5 basis points for both threshold scenarios and both graph representations. Especially with respect to *AP*, a clear decline for both graph formalisms can be observed. This effect can also be observed in the recall-precision curve in Fig. 7.10.

Finally, in Table 7.17 and Fig. 7.11 we examine the proposed quadtree segmentation approach on the BOT manuscript. Similar to AK, we observe rather low speed-up factors of 4 to 6. A reason for this observation might

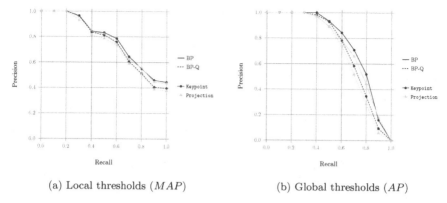

(a) Local thresholds (MAP)   (b) Global thresholds (AP)

Fig. 7.10: AK: Recall-precision curves using BP and BP-Q graph-matching algorithms.

Table 7.17: BOT: MAP, AP, and SF of the KWS system using different graph representations and BP in conjunction with quadtree segmentations. With ± we indicate the relative percental gain or loss in the accuracy of BP-Q when compared with BP. The best result per column is shown in bold.

Method	MAP	±	AP	±	SF
		BP			
Keypoint	45.06		33.84		
Projection	**49.57**		**38.33**		
		BP-Q			
Keypoint	39.58	−5.48	30.47	−3.37	4
Projection	47.24	−2.33	35.39	−2.94	6

be the rather large overlapping factor for `Keypoint` graphs ($o = 0.14$) and the high recursion depth for `Projection` graphs ($r = 3$). Moreover, in all cases, a general decrease of the accuracy can be observed.

Overall, we can conclude that BP-Q generally allows substantive speed-ups of the KWS process. That is, BP-Q is between 4 to 22 times faster when compared to BP, depending on the overlapping factor o and recursion depth r. However, these performance improvements with respect to the runtime are accompanied by a general loss of accuracy of up to 5 basis points for both local and global thresholds.

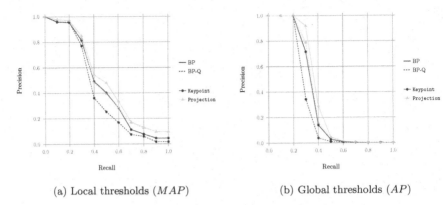

(a) Local thresholds (MAP)   (b) Global thresholds (AP)

Fig. 7.11: BOT: Recall-precision curves using BP and BP-Q graph-matching algorithms.

7.4.2   *KWS System Based on BP with Fast Rejection Methods*

The second group of speed-up approaches is based on two fast rejection methods, namely BP-FRN and BP-FRE (introduced in Section 6.3.2). That is, a pair of graphs is first examined by means of the node or edge distribution in a polar coordinate system (denoted by PGD-Node and PGD-Edge, respectively). BP is only employed if these distributions are similar enough (that is, if the resulting distance is below an additional threshold).

İn Table 7.18, we observe FR between 60% and 90% for BP-FRN in the GW manuscript, and between 85% and 95% for BP-FRE (depending on the graph representation). Hence, only 5% to 40% of all comparisons have to be carried out by the computationally expensive BP graph matching algorithm. Due to this filtering, we observe speed-up factors of 3 to 24 when compared with the original framework. Simultaneously both methods BP-FRN and BP-FRE achieve better accuracy levels than the plain method BP (see also Fig. 7.12). When compared with BP-FRN, BP-FRE generally leads to higher filter rates as well as higher accuracy levels in both threshold scenarios. In particular, BP-FRE in combination with Keypoint graphs lead to accuracy improvements of up to 4 basis points (when compared with BP on the same graphs).

On the PAR manuscript, accuracy improvements similar to those on GW can be observed for both BP-FRN and BP-FRE, as shown in Table 7.19. The achieved filter rates are between 74% and 94% for BP-FRN, and between 85% and 96% for BP-FRE (depending on the graph representation). This leads to speed-ups with factors between 4 to 16 when compared with the baseline system. Similar to GW, KWS accuracy can

Table 7.18: GW: *MAP*, *AP*, *FR*, and *SF* of the KWS system using different graph representations and fast rejection methods BP-FRN and BP-FRE. With ± we indicate the relative percental gain or loss in the accuracy of BP-FR when compared with BP. The best result per column is shown in bold.

Method	MAP	±	AP	±	FR	SF
			BP			
Keypoint	66.08		54.37		0.00	
Grid	60.02		45.52		0.00	
Projection	61.43		48.45		0.00	
Split	60.23		48.41		0.00	
			BP-FRN			
Keypoint	69.81	+3.73	56.45	+2.08	86.97	8
Grid	62.85	+2.83	46.39	+0.86	90.92	11
Projection	65.20	+3.77	51.19	+2.75	93.02	14
Split	63.15	+2.92	51.27	+2.86	58.17	3
			BP-FRE			
Keypoint	**70.61**	+4.52	**57.04**	+2.67	95.32	21
Grid	62.86	+2.84	45.20	−0.32	95.78	24
Projection	65.51	+4.08	50.69	+2.25	94.79	19
Split	64.01	+3.78	51.78	+3.37	85.38	7

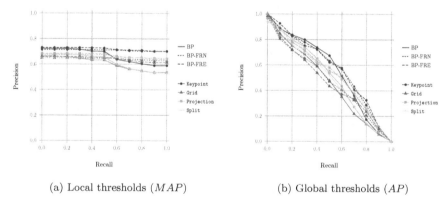

(a) Local thresholds (MAP)   (b) Global thresholds (AP)

Fig. 7.12: GW: Recall-precision curves using BP, BP-FRN, and BP-FRE graph-matching algorithms.

also be generally improved (see Fig. 7.13). That is, for both filtering methods the accuracy for local thresholds can be improved by between 5 and 7 basis points, while the accuracy for global thresholds can be improved by up to 16 basis points.

Table 7.19: PAR: MAP, AP, FR, and SF of the KWS system using different graph representations and fast rejection methods BP-FRN and BP-FRE. With $\pm$ we indicate the relative percental gain or loss in the accuracy of BP-FR when compared with BP. The best result per column is shown in bold.

Method	MAP	$\pm$	AP	$\pm$	FR	SF
			BP			
Keypoint	62.04		61.36		0.00	
Grid	56.50		43.55		0.00	
Projection	66.23		64.23		0.00	
Split	59.44		58.29		0.00	
			BP-FRN			
Keypoint	67.28	+5.24	67.04	+5.68	73.60	4
Grid	62.33	+5.83	59.52	+15.97	86.13	7
Projection	71.09	+4.86	**71.77**	+7.54	93.85	16
Split	65.43	+5.99	66.80	+8.51	82.84	6
			BP-FRE			
Keypoint	68.16	+6.12	67.42	+6.06	92.53	13
Grid	63.12	+6.62	59.85	+16.30	91.61	12
Projection	**72.03**	+5.80	71.48	+7.25	84.20	6
Split	66.49	+7.05	65.73	+7.44	92.69	14

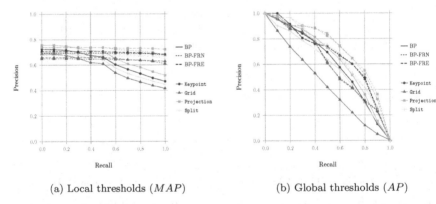

(a) Local thresholds (MAP)   (b) Global thresholds (AP)

Fig. 7.13: PAR: Recall-precision curves using BP, BP-FRN, and BP-FRE graph-matching algorithms.

Similar to GW and PAR, filter rates of 85% to 92% can be observed for both filter methods on the AK manuscript, as shown in Table 7.20. These filter rates lead to speed-up factors between 7 and 13 when compared with

Table 7.20: AK: *MAP*, *AP*, *FR*, and *SF* of the KWS system using different graph representations and fast rejection methods BP-FRN and BP-FRE. With ± we indicate the relative percental gain or loss in the accuracy of BP-FR when compared with BP. The best result per column is shown in bold.

Method	MAP	±	AP	±	FR	SF
			BP			
Keypoint	77.24		76.32		0.00	
Projection	76.02		74.38		0.00	
			BP-FRN			
Keypoint	**81.51**	+4.27	79.01	+2.69	85.45	7
Projection	79.09	+3.07	77.20	+2.82	91.09	11
			BP-FRE			
Keypoint	**81.51**	+4.27	**79.91**	+3.59	88.57	9
Projection	**81.51**	+5.49	79.26	+4.88	92.48	13

the original KWS framework. Yet, in contrast to the other manuscripts, the improvements with respect to accuracy are slightly lower and lie between 3 and 4 basis points for BP-FRN, and between 4 and 5 basis points for BP-FRE. These small improvements can also be observed in the recall-precision curves in Fig. 7.14.

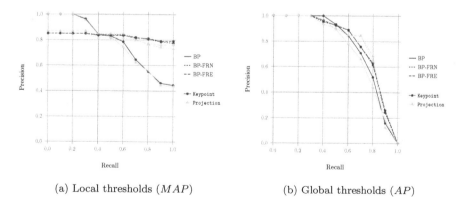

(a) Local thresholds (*MAP*)   (b) Global thresholds (*AP*)

Fig. 7.14: AK: Recall-precision curves using BP, BP-FRN, and BP-FRE graph-matching algorithms.

Table 7.21: BOT: MAP, AP, FR, and SF of the KWS system using different graph representations and fast rejection methods BP-FRN and BP-FRE. With ± we indicate the relative percental gain or loss in the accuracy of BP-FR when compared with BP. The best result per column is shown in bold.

Method	MAP	±	AP	±	FR	SF
BP						
Keypoint	45.06		33.84		0.00	
Projection	49.57		38.33		0.00	
BP-FRN						
Keypoint	56.10	+11.04	39.85	+6.01	85.64	7
Projection	53.77	+4.20	37.75	−0.58	84.20	6
BP-FRE						
Keypoint	**57.14**	+12.08	**40.48**	+6.64	84.75	7
Projection	52.56	+2.99	36.91	−1.42	86.20	7

On the last and most challenging manuscript, BOT, we observe rather low filter rates of 84% to 86%, leading to speed-up factors of 6 to 7, as shown in Table 7.21. Regarding KWS accuracy, we observe improvements for Keypoint graphs of 6 to 12 basis points when compared with BP. On Projection graphs, KWS accuracy is improved with respect to MAP only, while a small decline of the accuracy is observed for AP. These effects can be also observed in the recall-precision curves in Fig. 7.15.

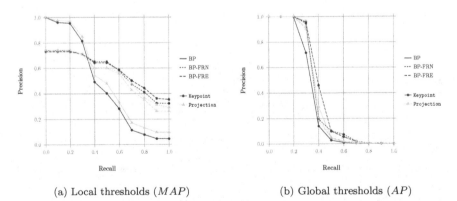

(a) Local thresholds (MAP)   (b) Global thresholds (AP)

Fig. 7.15: BOT: Recall-precision curves using BP, BP-FRN, and BP-FRE graph-matching algorithms.

We conclude that the proposed filter methods BP-FRN and BP-FRE allow substantial speed-ups when compared to the original KWS framework that relies on BP only. Moreover, the proposed filters are able to improve KWS accuracy in most scenarios. Hence, the advantage of BP-FRN and BP-FRE is twofold. It makes the graph-based KWS framework faster *and* more accurate. This makes this particular extension of the baseline system beneficial, especially in comparison with BP-Q, where the runtime is reduced but accuracy is not improved.

7.4.3 *KWS System Based on PGD*

Given the high accuracies achieved by BP-FRN and BP-FRE, one might wonder how the linear time graph dissimilarity *polar graph dissimilarity* (PGD) (introduced in Section 5.4) performs as an isolated graph dissimilarity measure for KWS, rather than as a filter method for BP. In Tables 7.22 and Fig. 7.16, the KWS accuracies for PGD with node-based histograms (PGD-Node) as well as edge-based histograms (PGD-Edge) are

Table 7.22: GW: MAP, AP, and SF of the KWS system using different graph representations and PGD with nodes (PGD-Node) and edges (PGD-Edge). With $\pm$ we indicate the relative percental gain or loss in the accuracy of PGD when compared with BP. The best result per column is shown in bold.

Method	MAP	$\pm$	AP	$\pm$	SF
		BP			
Keypoint	66.08		54.37		
Grid	60.02		45.52		
Projection	61.43		48.45		
Split	60.23		48.41		
		PGD-Node			
Keypoint	58.47	−7.62	47.13	−7.24	1,058
Grid	54.32	−5.70	41.19	−4.34	2,011
Projection	58.13	−3.30	45.99	−2.46	1,093
Split	53.59	−6.64	40.71	−7.70	1,133
		PGD-Edge			
Keypoint	**68.78**	+2.70	**58.10**	+3.73	767
Grid	64.92	+4.89	52.99	+7.47	1,077
Projection	54.71	−6.72	44.83	−3.61	1,000
Split	63.04	+2.81	49.46	+1.05	808

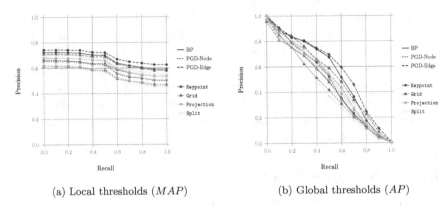

(a) Local thresholds (MAP)   (b) Global thresholds (AP)

Fig. 7.16: GW: Recall-precision curves using BP, PGD-Node, and PGD-Edge graph-matching algorithms.

compared against BP on the GW manuscript. We observe that PGD-Node results in a lower KWS accuracy rate when compared with BP, whereas PGD-Edge performs similarly to or even better than BP. This is particularly impressive as the resulting speed-up factors of this system are between 770 and 2,010 when compared with the cubic time algorithm BP.

On the PAR manuscript, similar speed-up factors as in GW can be observed (shown in Table 7.23 and Fig. 7.17). However, these speed-ups are accompanied by a rather large decline in the accuracy in the case of PGD-Node. In contrast to this, PGD-Edge achieves a similar level of accuracy as the baseline system BP. In particular, PGD-Edge in combination with Split graphs improves the accuracy for both threshold scenarios. On the other hand, PGD-Edge does not perform very well on the Projection graphs.

Similar to the previous manuscripts, clear speed-up factors of 310 to 1,060 can be registered on the AK manuscript, as shown in Table 7.24. Yet, this improvement regarding runtime is accompanied by a general loss in accuracy in the case of PGD-Node. In particular, on Keypoint graphs a decline of 8 to 11 basis points has to be observed. In contrast to this, PGD-Edge is generally able to keep up with or even outperform the original BP framework. The same effects can be observed in the recall-precision curves in Fig. 7.18.

Finally, in Table 7.25 and Fig. 7.19 the proposed linear time graph dissimilarities PGD-Node and PGD-Edge are evaluated in the challenging BOT manuscript. In this dataset we observe very high speed-up factors of 965 to 5,260. Simultaneously, accuracies similar to those with BP can be

Table 7.23: PAR: MAP, AP, and SF of the KWS system using different graph representations and PGD with nodes (PGD-Node) and edges (PGD-Edge). With ± we indicate the relative percental gain or loss in the accuracy of PGD when compared with BP. The best result per column is shown in bold.

Method	MAP	±	AP	±	SF
		BP			
Keypoint	62.04		61.36		
Grid	56.50		43.55		
Projection	66.23		64.23		
Split	59.44		58.29		
		PGD-Node			
Keypoint	46.71	−15.33	36.30	−25.06	896
Grid	31.80	−24.70	28.08	−15.47	2,487
Projection	50.05	−16.18	45.91	−18.32	1,325
Split	45.02	−14.42	31.46	−26.83	1,076
		PGD-Edge			
Keypoint	61.47	−0.57	67.20	+5.84	877
Grid	53.53	−2.97	52.49	+8.94	1,302
Projection	57.51	−8.72	58.03	−6.20	1,060
Split	**62.62**	+3.18	**71.23**	+12.94	868

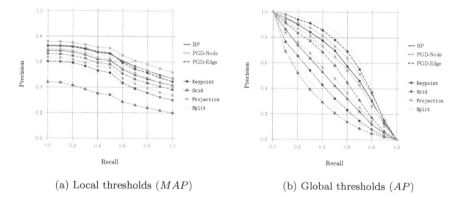

(a) Local thresholds (MAP) (b) Global thresholds (AP)

Fig. 7.17: PAR: Recall-precision curves using BP, PGD-Node, and PGD-Edge graph-matching algorithms.

achieved. Using PGD-Node rather than BP, a small decline in accuracy can be observed on `Projection` graphs, while on `Keypoint` graphs the accuracy is slightly improved. Substantial improvements can be observed

Table 7.24: AK: MAP, AP, and SF of the KWS system using different graph representations and PGD with nodes (PGD-Node) and edges (PGD-Edge). With ± we indicate the relative percental gain or loss in the accuracy of PGD when compared with BP. The best result per column is shown in bold.

Method	MAP	±	AP	±	SF
BP					
Keypoint	**77.24**		**76.32**		
Projection	76.02		74.38		
PGD-Node					
Keypoint	69.66	−7.58	65.40	−10.92	1,062
Projection	72.36	−3.66	69.52	−4.86	418
PGD-Edge					
Keypoint	76.17	−1.07	74.76	−1.56	951
Projection	76.98	+0.96	74.70	+0.32	310

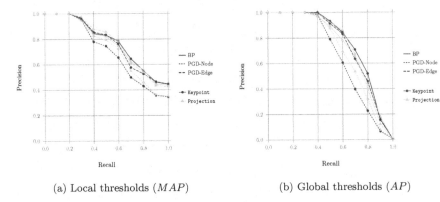

(a) Local thresholds (MAP) (b) Global thresholds (AP)

Fig. 7.18: AK: Recall-precision curves using BP, PGD-Node, and PGD-Edge graph-matching algorithms.

for `Keypoint` graphs by means of PGD-Edge, with respect to both runtime and accuracy. That is, the accuracy can be improved for both threshold scenarios by 7 basis points.

To conclude this section, we observe that PGD-Node generally leads to large speed-ups when compared with BP. However, a decline in KWS accuracy for local and global thresholds needs to be taken into account. In contrast to this, PGD-Edge allows similar speed-up factors and is also able to

Table 7.25: BOT: MAP, AP, and SF of the KWS system using different graph representations and PGD with nodes (PGD-Node) and edges (PGD-Edge). With $\pm$ we indicate the relative percental gain or loss in the accuracy of PGD when compared with BP. The best result per column is shown in bold.

Method	MAP	$\pm$	AP	$\pm$	SF
BP					
Keypoint	45.06		33.84		
Projection	49.57		38.33		
PGD-Node					
Keypoint	45.81	+0.75	33.93	+0.09	4,051
Projection	46.13	−3.44	35.48	−2.85	965
PGD-Edge					
Keypoint	**52.11**	+7.05	**41.21**	+7.37	5,260
Projection	49.53	−0.04	38.41	+0.08	1,065

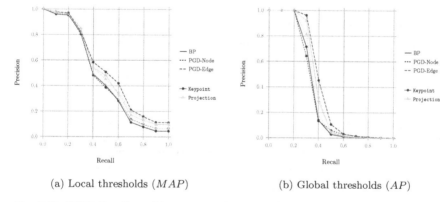

(a) Local thresholds (MAP)   (b) Global thresholds (AP)

Fig. 7.19: BOT: Recall-precision curves using BP, PGD-Node, and PGD-Edge graph-matching algorithms.

keep up with or even outperform the cubic time algorithm BP with respect to KWS accuracy. We can also observe that both methods (i.e. PGD-Node and PGD-Edge) have a beneficial impact when employed as filter methods in conjunction with BP. That is, BP-FRN and BP-FRE achieve better results in most cases than the isolated graph dissimilarities BP and PGD.

7.5 KWS System Based on HED

In this section, a suboptimal algorithm for GED with quadratic time complexity, namely *Hausdorff edit distance* (HED), is evaluated [79] (described in Section 5.3.3). This particular algorithm offers a lower bound for the GED and provides no valid edit path in contrast to BP. Due to its lower time complexity, HED is able to consider a larger node context during the assignment optimisation. Thus, the so-called *context-aware Hausdorff edit distance* (CED) (see Subsection 5.3.3.1) can be applied, which is potentially able to reduce the approximation error with respect to exact GED.

We evaluate HED in a first series of experiments in Subsection 7.5.1. Then, CED is evaluated by means of three different context sizes in Subsection 7.5.2.

7.5.1 *Hausdorff Edit Distance*

The KWS results achieved with *Hausdorff edit distance* (HED) on the GW manuscript are shown in Table 7.26 and Fig. 7.20. HED is able to improve the accuracy of BP by 3 to 5 basis points with respect to MAP, and by 3 to 8 basis points with respect to AP. Unlike BP, HED allows multiple assignments among substructures of graphs. We assume that this property is beneficial in the context of handwriting because it allows a kind of "warping" between characters of different size and style. Moreover, HED allows

Table 7.26: GW: MAP, AP, and SF of the KWS system using different graph representations and HED. With ± we indicate the relative percental gain or loss in the accuracy of HED when compared with BP. The best result per column is shown in bold.

Method	MAP	±	AP	±	SF
		BP			
Keypoint	66.08		54.37		
Grid	60.02		45.52		
Projection	61.43		48.45		
Split	60.23		48.41		
		HED			
Keypoint	**69.28**	+3.19	**57.38**	+3.01	95
Grid	62.78	+2.75	51.27	+5.75	116
Projection	66.71	+5.28	56.10	+7.65	88
Split	65.12	+4.89	54.32	+5.91	108

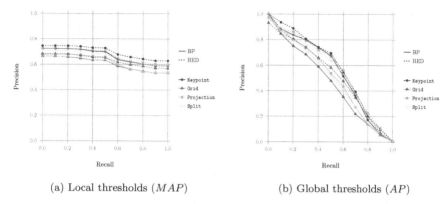

(a) Local thresholds (*MAP*)   (b) Global thresholds (*AP*)

Fig. 7.20: GW: Recall-precision curves using BP and HED graph-matching algorithms.

us to reduce the computational complexity of the KWS process leading to speed-up factors of 88 to 116 when compared with the baseline framework BP.

Table 7.27: PAR: MAP, AP, and SF of the KWS system using different graph representations and HED. With $\pm$ we indicate the relative percental gain or loss in the accuracy of HED when compared with BP. The best result per column is shown in bold.

Method	MAP	$\pm$	AP	$\pm$	SF
		BP			
Keypoint	62.04		61.36		
Grid	56.50		43.55		
Projection	66.23		64.23		
Split	59.44		58.29		
		HED			
Keypoint	69.23	+7.19	74.23	+12.87	84
Grid	60.74	+4.24	58.47	+14.92	112
Projection	**72.82**	+6.59	**76.69**	+12.46	100
Split	72.79	+13.35	73.93	+15.64	79

Next, HED is evaluated on the PAR manuscript, as shown in Table 7.27. We observe large improvements in the accuracy of 4 to 13 basis points for local thresholds, and of 13 to 16 basis points for global thresholds. These results emphasise the usefulness of multiple assignments of graph substructures provided by HED. The effects of this improvement can also be observed on the recall-precision curves shown in Fig. 7.21. PAR is similar to GW, in that we can observe speed-up factors of 79 to 112 when compared with the BP matching algorithm in PAR.

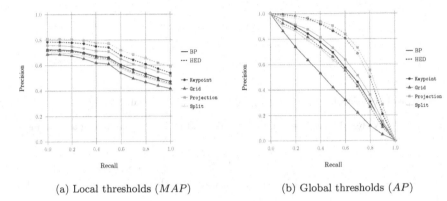

(a) Local thresholds (MAP)   (b) Global thresholds (AP)

Fig. 7.21: PAR: Recall-precision curves using BP and HED graph-matching algorithms.

KWS using HED also improves accuracy in the AK manuscript, as shown in Table 7.28 and Fig. 7.22. That is, for both threshold scenarios, accuracy can be improved by 2 and 5 basis points for `Keypoint` and `Projection` graphs, respectively. Regarding the speed-up factors, we observe that HED is between between 55 and 91 times faster than BP.

Table 7.28: AK: MAP, AP, and SF of the KWS system using different graph representations and HED. With ± we indicate the relative percental gain or loss in the accuracy of HED when compared with BP. The best result per column is shown in bold.

Method	MAP	±	AP	±	SF
		BP			
Keypoint	77.24		76.32		
Projection	76.02		74.38		
		HED			
Keypoint	79.72	+2.48	78.61	+2.29	91
Projection	**81.06**	+5.04	**79.81**	+5.43	55

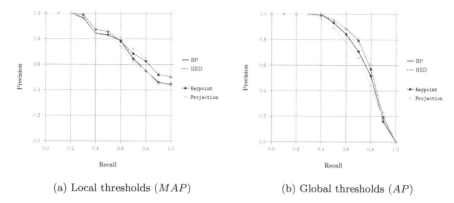

(a) Local thresholds (*MAP*)   (b) Global thresholds (*AP*)

Fig. 7.22: AK: Recall-precision curves using BP and HED graph-matching algorithms.

Finally, HED is examined on the BOT manuscript, as shown in Table 7.29 and Fig. 7.23. On `Keypoint` graphs, we observe improvements of 7 and 6 basis points with respect to local and global thresholds, respectively. Slightly smaller improvements of about 2 and 1 basis points can be observed for the `Projection` graphs. In terms of computation time, we register speed-up factors of 91 to 230 when compared with the baseline framework BP.

We can conclude that using HED in our KWS framework offers two crucial advantages. First, accuracy can be improved for both threshold scenarios. This might be due to the multiple node assignments allowed in HED, which in turn allows a certain degree of warping when compared

Table 7.29: BOT: *MAP*, *AP*, and *SF* of the KWS system using different graph representations and HED. With ⊥ we indicate the relative percental gain or loss in the accuracy of HED when compared with BP. The best result per column is shown in bold.

Method	*MAP*	±	*AP*	±	*SF*
		BP			
`Keypoint`	45.06		33.84		
`Projection`	49.57		38.33		
		HED			
`Keypoint`	51.69	+6.63	**40.29**	+6.45	230
`Projection`	**51.74**	+2.17	38.89	+0.56	91

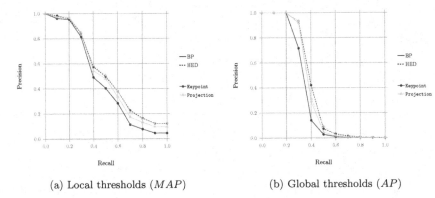

(a) Local thresholds (MAP) (b) Global thresholds (AP)

Fig. 7.23: BOT: Recall-precision curves using BP and HED graph-matching algorithms.

Table 7.30: GW: MAP, AP, and SF of the KWS system using different graph representations and CED with different context radii n. With $\pm$ we indicate the relative percental gain or loss in the accuracy of CED when compared with BP. The best result per column is shown in bold.

Method	MAP	$\pm$	AP	$\pm$	SF
		BP			
Keypoint	66.08		54.37		
Grid	60.02		45.52		
Projection	61.43		48.45		
Split	60.23		48.41		
		CED ($n = 2$)			
Keypoint	**71.23**	+5.15	**59.15**	+4.78	41
Grid	66.90	+6.87	52.70	+7.18	53
Projection	67.04	+5.62	54.32	+5.87	49
Split	66.59	+6.36	55.83	+7.42	46
		CED ($n = 4$)			
Keypoint	68.46	+2.37	57.82	+3.45	21
Grid	64.06	+4.04	51.34	+5.82	27
Projection	64.51	+3.08	54.30	+5.85	27
Split	64.15	+3.92	55.08	+6.67	28
		CED ($n = 6$)			
Keypoint	65.23	−0.85	55.24	+0.87	17
Grid	63.24	+3.21	49.64	+4.12	18
Projection	60.25	−1.18	48.69	+0.25	18
Split	62.48	+2.25	53.22	+4.81	19

to BP-based approaches. Second, the quadratic time complexity of HED offers speed-up factors between 55 and 230 when compared with the cubic time algorithm BP.

7.5.2 *Context-aware Hausdorff Edit Distance*

In this subsection, we evaluate our KWS framework in conjunction with *context-aware Hausdorff edit distance* (CED), an extension of HED that takes a larger node context into account — rather than adjacent edges only. Using CED rather than HED might have a positive and a negative effect, namely smaller approximation errors but also higher runtimes.

In Table 7.30, we compare CED with three different context radii $n = \{2, 4, 6\}$ in the GW manuscript. We observe that a context radius of $n = 2$ is ideal for all graph representations of this dataset. This effect can also be observed in the recall-precision curves in Fig. 7.24. We observe improvements of 5 to 7 basis points when compared to BP for both threshold scenarios. Regarding the speed-up factors, we generally observe a decline of the speed-up when n is increased. Clearly, this is due to the higher number of node assignments that have to be carried out in the larger node context. Nevertheless, we observe speed-up factors between 17 and 53 for all tested values for n when compared with BP.

Slightly different observations can be made about the PAR manuscript, as shown in Table 7.31 and Fig. 7.25. That is, smaller values for n seem to be beneficial for two representations (i.e. Keypoint and Projection), while

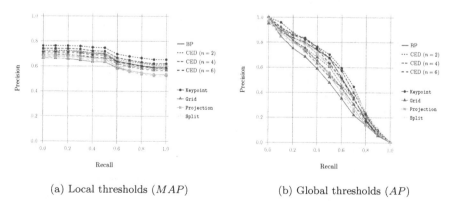

(a) Local thresholds (MAP)   (b) Global thresholds (AP)

Fig. 7.24: GW: Recall-precision curves using BP and CED graph-matching algorithms.

Table 7.31: PAR: MAP, AP, and SF of the KWS system using different graph representations and CED with different context radii n. With ± we indicate the relative percental gain or loss in the accuracy of CED when compared with BP. The best result per column is shown in bold.

Method	MAP	±	AP	±	SF
		BP			
Keypoint	62.04		61.36		
Grid	56.50		43.55		
Projection	66.23		64.23		
Split	59.44		58.29		
		CED ($n = 2$)			
Keypoint	68.39	+6.35	73.13	+11.77	35
Grid	63.06	+6.56	56.35	+12.80	54
Projection	**74.10**	+7.87	73.44	+9.21	57
Split	70.86	+11.42	74.69	+16.40	44
		CED ($n = 4$)			
Keypoint	68.64	+6.60	71.78	+10.42	22
Grid	65.09	+8.59	61.12	+17.57	29
Projection	73.99	+7.76	73.82	+9.59	28
Split	72.91	+13.47	**77.94**	+19.65	22
		CED ($n = 6$)			
Keypoint	67.19	+5.15	70.02	+8.66	14
Grid	65.98	+9.48	58.92	+15.37	18
Projection	72.45	+6.22	71.63	+7.40	18
Split	73.60	+14.16	75.74	+17.45	15

Grid and Split graphs take advantage of the larger node context. Overall, we observe accuracy improvements of 5 to 14 basis points with respect to MAP, and of 7 to 20 basis points with respect to AP. Moreover, we notice speed-up factors between 14 and 57 when compared with BP.

In the AK manuscript, we observe that $n = 2$ is optimal for Projection graphs, while $n = 6$ is optimal for Keypoint graphs, as shown in Table 7.32. Accuracy improvements of 5 and 6 basis points can be noticed for Keypoint and Projection graphs, respectively. These improvements can also be observed in the recall-precision curves in Fig. 7.26. Regarding the computation time, we observe rather low speed-up factors between 8 and 44. The reason for these rather small improvements might be the high level of connectivity of the nodes in this dataset, i.e. the large number of adjacent

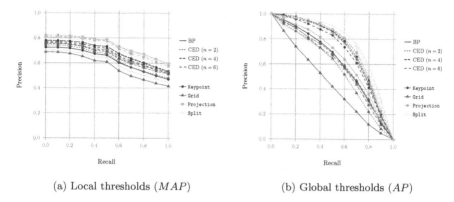

(a) Local thresholds (*MAP*)　　　　　(b) Global thresholds (*AP*)

Fig. 7.25: PAR: Recall-precision curves using BP and CED graph-matching algorithms.

Table 7.32: AK: *MAP*, *AP*, and *SF* of the KWS system using different graph representations and CED with different context radii n. With ± we indicate the relative percental gain or loss in the accuracy of CED when compared with BP. The best result per column is shown in bold.

Method	*MAP*	±	*AP*	±	*SF*
		BP			
Keypoint	77.24		76.32		
Projection	76.02		74.38		
		CED ($n = 2$)			
Keypoint	80.73	+3.49	79.07	+2.75	44
Projection	80.28	+4.26	78.12	+3.74	26
		CED ($n = 4$)			
Keypoint	81.43	+4.19	80.50	+4.18	20
Projection	82.42	+6.40	80.85	+6.47	12
		CED ($n = 6$)			
Keypoint	**82.69**	+5.45	**81.46**	+5.14	12
Projection	81.58	+5.56	79.67	+5.29	8

edges, which leads to a substantial increase in the computational burden for larger n.

Finally, we evaluate CED on the BOT manuscripts, as shown in Table 7.33 and Fig. 7.27. We observe only marginal accuracy improve

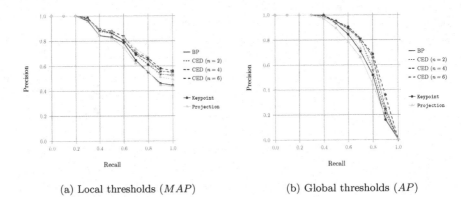

(a) Local thresholds (MAP)　　　　　　(b) Global thresholds (AP)

Fig. 7.26: AK: Recall-precision curves using BP and CED graph-matching algorithms.

Table 7.33: BOT: MAP, AP, and SF of the KWS system using different graph representations and CED with different context radii n. With $\pm$ we indicate the relative percental gain or loss in the accuracy of CED when compared with BP. The best result per column is shown in bold.

Method	MAP	$\pm$	AP	$\pm$	SF
		BP			
Keypoint	45.06		33.84		
Projection	49.57		38.33		
		CED ($n = 2$)			
Keypoint	51.59	+6.53	**40.34**	+6.50	118
Projection	51.98	+2.41	39.52	+1.19	46
		CED ($n = 4$)			
Keypoint	52.18	+7.12	39.21	+5.37	55
Projection	52.02	+2.45	38.24	−0.09	22
		CED ($n = 6$)			
Keypoint	**52.45**	+7.39	40.20	+6.36	36
Projection	50.59	+1.02	39.03	+0.70	16

ments of up to 2 basis points for local thresholds, and of about 1 basis point for global thresholds using the `Projection` graphs. In contrast to this, we observe substantial accuracy improvements of 7 basis points for local thresholds, and of 5 to 7 basis points for global thresholds using the graph

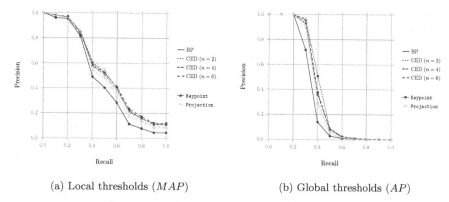

(a) Local thresholds (*MAP*)   (b) Global thresholds (*AP*)

Fig. 7.27: BOT: Recall-precision curves using BP and CED graph-matching algorithms.

formalism `Keypoint`. In terms of computation time, we observe speed-up factors between 16 and 118 when compared with the basic algorithm BP.

Overall, we observe clear accuracy and performance improvements in CED when compared with the baseline algorithm BP. In particular, the larger node context seems to be beneficial for all graph representations and manuscripts, leading to a general boost in accuracy. Moreover, CED does not only allow very high accuracy rates, but also offers a good compromise in terms of runtime due to the user-defined context radius n.

7.6   KWS System Based on BP2

In this section, we examine a combination of the two previous suboptimal algorithms for GED, namely BP and HED. The so-called *bipartite graph edit distance 2* (BP2) (described in Section 5.3.4) is similar to BP: an upper bound of GED resulting in a valid edit-path. However, BP2 offers quadratic time complexity — given by the assignment of HED — in contrast to the cubic time assignment used in BP.

On the GW manuscript, we observe MAP improvements on all graph representations but `Projection` (here a small decline of 0.6 basis points is observed) (see Table 7.34 and Fig. 7.28). BP2, in conjunction with `Keypoint` graphs, achieves the overall best result with regard to MAP. With respect to AP we observe that BP2 is able to improve the results of BP on two out of four graph representations. However, the overall best result with respect to AP is still achieved by BP (on `Keypoint` graphs). Overall, we note very similar accuracy rates compared with BP. However,

Table 7.34: GW: *MAP*, *AP*, and *SF* of the KWS system using different graph representations and BP2. With ± we indicate the relative percental gain or loss in the accuracy of BP2 when compared with BP. The best result per column is shown in bold.

Method	*MAP*	±	*AP*	±	*SF*
		BP			
Keypoint	66.08		**54.37**		
Grid	60.02		45.52		
Projection	61.43		48.45		
Split	60.23		48.41		
		BP2			
Keypoint	**68.42**	+2.33	52.47	−1.90	91
Grid	62.10	+2.07	48.35	+2.82	114
Projection	60.83	−0.60	48.25	−0.19	100
Split	64.24	+4.02	51.58	+3.17	122

the quadratic time behaviour of BP2 allows speed-up factors between 91 and 122 when compared with BP.

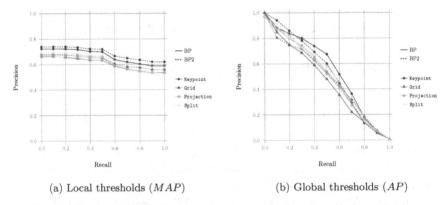

(a) Local thresholds (*MAP*)   (b) Global thresholds (*AP*)

Fig. 7.28: GW: Recall-precision curves using BP and BP2 graph-matching algorithms.

On the second manuscript, PAR, we observe a clear decline of 7 and 4 basis points for local and global thresholds, respectively (see Table 7.35) in the accuracy of BP2 on `Keypoint` graphs. The same applies to `Projection` with respect to local thresholds. These effects are also visible in the recall-precision curves in Fig. 7.29. However, on both `Grid` and `Split` graphs

Table 7.35: PAR: MAP, AP, and SF of the KWS system using different graph representations and BP2. With ± we indicate the relative percental gain or loss in the accuracy of BP2 when compared with BP. The best result per column is shown in bold.

Method	MAP	±	AP	±	SF
		BP			
Keypoint	62.04		61.36		
Grid	56.50		43.55		
Projection	66.23		64.23		
Split	59.44		58.29		
		BP2			
Keypoint	55.03	−7.01	57.69	−3.67	78
Grid	57.00	+0.50	57.96	+14.41	118
Projection	63.35	−2.88	65.53	+1.30	106
Split	**68.69**	+9.25	**69.54**	+11.25	93

BP2 achieves accuracy improvements of more than 14 basis points. BP2 in conjunction with `Split` graphs achieves the overall best accuracy regarding both MAP and AP. Similar to GW, we observe speed-up factors between 78 and 118 when compared with BP.

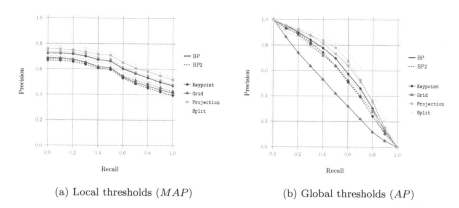

(a) Local thresholds (MAP) (b) Global thresholds (AP)

Fig. 7.29: PAR: Recall-precision curves using BP and BP2 graph-matching algorithms.

In Table 7.36, we observe a decline in accuracy for all graph representations and threshold scenarios in the AK manuscript. However, the accuracy levels achieved by BP2 are still comparable to those of BP. This can also be

Table 7.36: AK: MAP, AP, and SF of the KWS system using different graph representations and BP2. With $\pm$ we indicate the relative percental gain or loss in the accuracy of BP2 when compared with BP. The best result per column is shown in bold.

Method	MAP	$\pm$	AP	$\pm$	SF
		BP.			
Keypoint	**77.24**		**76.32**		
Projection	76.02		74.38		
		BP2			
Keypoint	74.86	−2.38	73.61	−2.71	92
Projection	75.46	−0.56	73.85	−0.53	56

observed in the recall-precision curves in Fig. 7.30. Moreover, we account for substantial speed-up factors of 92 and 56 on `Keypoint` and `Projection` graphs, respectively.

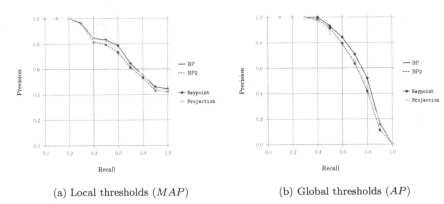

(a) Local thresholds (MAP)   (b) Global thresholds (AP)

Fig. 7.30: AK: Recall-precision curves using BP and BP2 graph-matching algorithms.

Finally, we compare BP2 with BP in the BOT manuscript, as shown in Table 7.37 and Fig. 7.31. We observe improvements in accuracy for both graph representations and threshold scenarios. Moreover, we observe substantial speed-up factors of between 110 and 279 when compared to the baseline framework BP.

We conclude that BP2 achieves on all manuscripts but AK and for both scenarios (MAP and AP) has similar or even better results than BP. BP2

Table 7.37: BOT: MAP, AP, and SF of the KWS system using different graph representations and BP2. With $\pm$ we indicate the relative percental gain or loss in the accuracy of BP2 when compared with BP. The best result per column is shown in bold.

Method	MAP	$\pm$	AP	$\pm$	SF
		BP			
Keypoint	45.06		33.84		
Projection	49.57		38.33		
		BP2			
Keypoint	**50.94**	+5.88	38.20	+4.36	279
Projection	50.49	+0.92	**39.25**	+0.92	110

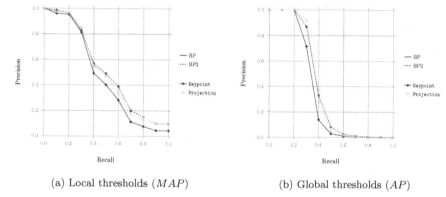

(a) Local thresholds (MAP)   (b) Global thresholds (AP)

Fig. 7.31: BOT: Recall-precision curves using BP and BP2 graph-matching algorithms.

also leads to a substantial speed-up of the complete KWS framework when compared to the baseline system BP.

7.7   Graph-Based Ensembles for KWS

In this section, different graph-based ensemble methods for KWS (introduced in Section 6.3.3) are compared. We make use of multiple graph representations at a time and condense the resulting graph edit distances by the use of different statistical methods. The rationale for this contribution is that ensemble methods often allow higher accuracy rates than single classification methods [299]. However, in our scenario, this advantage is accompanied by an increase in the computational complexity due to the requirement of multiple graph matchings.

Clearly, any dissimilarity framework presented so far could be used for the combination of distances. However, for the sake of convenience, we first present ensemble results achieved with the baseline system *bipartite graph edit distance* (BP) only. Second, results on the combination of the remaining graph matching algorithms (i.e. *Hausdorff edit distance* (HED), *context-aware Hausdorff edit distance* (CED), and *bipartite graph edit distance 2* (BP2)) are presented in an overview. Note that we compare the ensemble methods with the best results obtained by a single graph dissimilarity (i.e. BP, HED, CED, and BP2 as isolated graph matching algorithms). In the vast majority of cases, the underlying graph representations rely on Keypoint or Projection graphs.

7.7.1 *Ensemble Methods Based on BP*

Basically, we employ five different types of combination methods. That is, we condense several graph edit distances that are computed on different graph representations of the same pair of query and document words by either taking the minimum, maximum, or mean distance (denoted by min, max, and mean, respectively). Moreover, we employ two different weighted sums of the graph distances that are either based on a user-defined weighting parameter γ (denoted by sum$_\gamma$) or the MAP of the single methods (denoted by sum$_{\mathrm{map}}$). For further details regarding these ensemble methods we refer to Section 6.3.3.

In the first evaluation, we compare three ensemble methods (min, max, and mean) that are based on all four graph representations and on the two most promising graph formalisms only, namely Keypoint and Projection. Table 7.38 shows the results of all ensemble methods that are based on two or four graph representations. In three out of four cases the ensembles that use four rather than two graph representations achieve the best results (twice by means of the mean operation and once by means of the minimum operation). However, the ensembles that rely on two representations also achieve remarkable improvements when compared with the baseline system BP. Moreover, for two manuscripts, namely AK and BOT, only distances for the Keypoint and Projection graphs have been computed. Hence, we limit the more thorough evaluation of ensembles for KWS on the Keypoint and Projection graphs.

In Table 7.39 we present the results of all five ensemble techniques and compare these results with the best possible single system achieved with BP on the GW manuscript. We observe clear improvements in the accuracy

Table 7.38: GW and PAR: MAP and AP of the KWS system using different BP-based ensemble methods and two or four graph representations. With $\pm$ we indicate the relative percental gain or loss in the accuracy of the ensemble methods when compared with BP. The best result per column is shown in bold.

Method	GW				PAR			
	MAP	$\pm$	AP	$\pm$	MAP	$\pm$	AP	$\pm$
Single								
BP	66.08		54.37		66.23		64.23	
Ensemble (two representations)								
min	**71.09**	+5.01	58.86	+4.49	73.93	+7.70	77.17	+12.94
max	66.71	+0.63	53.98	−0.40	70.07	+3.84	75.04	+10.81
mean	70.23	+4.15	58.39	+4.02	76.70	+10.47	80.91	+16.68
Ensemble (four representations)								
min	70.56	+4.48	**59.15**	+4.78	67.90	+1.67	70.38	+6.15
max	62.58	−3.51	48.72	−5.65	67.57	+1.34	63.91	−0.32
mean	69.16	+3.08	59.00	+4.63	**79.38**	+13.15	**85.43**	+21.20

Table 7.39: GW: MAP and AP of the KWS system using different BP-based ensemble methods. With $\pm$ we indicate the relative percental gain or loss in the accuracy of the ensemble methods when compared with BP. The best result per column is shown in bold.

Method	MAP	$\pm$	AP	$\pm$
Single				
BP	66.08		54.37	
Ensemble				
min	**71.09**	+5.01	**58.86**	+4.49
max	66.71	+0.63	53.98	−0.40
mean	70.23	+4.15	58.39	+4.02
sum_γ	68.44	+2.35	56.97	+2.60
sum_{map}	70.20	+4.12	56.75	+2.38

rates of nine out of ten scenarios. In particular, the ensemble methods min and mean improve the accuracy rates by between 4 and 5 basis points for both threshold scenarios. In contrast to this, the ensemble method max achieves only a small improvement with respect to MAP or even a small decrease in accuracy with regard to AP.

Fig. 7.32 shows the recall-precision curves for all ensemble methods. Note the normalisation effect of local thresholds. That is, most keywords are retrieved completely (i.e. we observe an *AP* of 100%), while some keywords are not retrieved at all (i.e. an *AP* of 0% is achieved). Consequently a linear *MAP* can be observed over the whole recall range. We assume that this effect is due to certain tendencies of the ensemble methods towards minimisation or maximisation of GED. Hence the application of local thresholds can become more difficult where the GEDs become more similar. Similar effects can also be observed for the remaining ensemble methods.

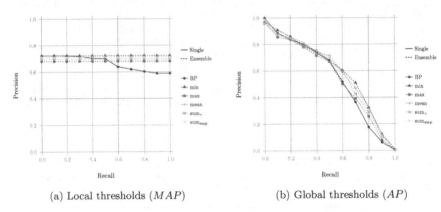

(a) Local thresholds (*MAP*)　　　　(b) Global thresholds (*AP*)

Fig. 7.32: GW: Recall-precision curves using BP and BP-based ensemble methods.

Quite similar observations can be made on the PAR manuscript, as shown in Table 7.40. In contrast to GW, however, the ensemble methods sum$_\gamma$ and mean rather than min lead to the largest accuracy improvements of 11 and 17 basis points for local and global thresholds, respectively. Overall, we see clear improvements in accuracy using all five ensemble methods when compared with the single method BP. These improvements can also be observed in the recall-precision curves in Fig. 7.33.

On the AK manuscript, we also observe similar accuracy improvements of about 7 basis points with respect to local and global thresholds by means of the ensemble methods mean, sum$_\gamma$, and sum$_{map}$ (see Table 7.41 and Fig. 7.34). Similarly to the two previous manuscripts, the ensemble method max leads to the smallest improvements when compared with the baseline system BP.

Table 7.40: PAR: MAP and AP of the KWS system using different BP-based ensemble methods. With ± we indicate the relative percental gain or loss in the accuracy of the ensemble methods when compared with BP. The best result per column is shown in bold.

Method	MAP	±	AP	±
		Single		
BP	66.23		64.23	
		Ensemble		
min	73.93	+7.70	77.17	+12.94
max	70.07	+3.84	75.04	+10.81
mean	76.70	+10.47	80.91	+16.68
sum$_\gamma$	74.51	+8.28	75.59	+11.36
sum$_{map}$	**76.80**	+10.57	**80.96**	+16.73

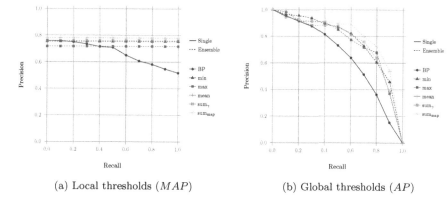

(a) Local thresholds (MAP)　　　　　(b) Global thresholds (AP)

Fig. 7.33: PAR: Recall-precision curves using BP and BP-based ensemble methods.

We also observe substantial improvements in the challenging BOT manuscript, as shown in Table 7.42. That is, we can report improvements of between 16 and 19 basis points with respect to local thresholds, and between 7 and 10 basis points with respect to global thresholds. In particular, the ensemble methods mean, sum$_\gamma$, and sum$_{map}$ offer massive improvements in both threshold scenarios. The improvement can also be observed in the recall-precision curves in Fig. 7.35.

We conclude that the proposed ensemble methods generally lead to clear accuracy improvements with respect to both MAP and AP. In particu-

Table 7.41: AK: MAP and AP of the KWS system using different BP-based ensemble methods. With ± we indicate the relative percental gain or loss in the accuracy of the ensemble methods when compared with BP. The best result per column is shown in bold.

Method	MAP	±	AP	±
	Single			
BP	77.24		76.32	
	Ensemble			
min	82.75	+5.51	80.29	+3.97
max	82.09	+4.85	80.56	+4.24
mean	84.25	+7.01	82.31	+5.99
sum_γ	**84.77**	+7.53	80.15	+3.83
sum_{map}	84.25	+7.01	**83.34**	+7.02

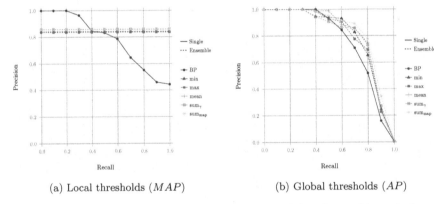

(a) Local thresholds (MAP)　　　　　(b) Global thresholds (AP)

Fig. 7.34: AK: Recall-precision curves using BP and BP-based ensemble methods.

lar, we observe improvements of up to 19 basis points in the challenging manuscript BOT when compared to the best result of BP. On most of the manuscripts and threshold scenarios, we register the largest improvements by means of the ensemble methods mean, sum_γ, or sum_{map}. Vice versa, the ensemble method max leads to the smallest improvements in accuracy when compared with the baseline system.

Note that ensemble methods generally lead to an increase of the computational complexity. In our particular experiment this means that we need to compute two graph dissimilarities rather than one (plus the computa-

Table 7.42: BOT: MAP and AP of the KWS system using different BP-based ensemble methods. With ± we indicate the relative percental gain or loss in the accuracy of the ensemble methods when compared with BP. The best result per column is shown in bold.

Method	MAP	±	AP	±
		Single		
BP	49.57		38.33	
		Ensemble		
min	65.19	+15.62	45.38	+7.05
max	67.57	+18.00	45.46	+7.13
mean	**68.88**	+19.31	48.18	+9.85
sum$_\gamma$	68.77	+19.20	**48.29**	+9.96
sum$_{\mathrm{map}}$	**68.88**	+19.31	48.28	+9.95

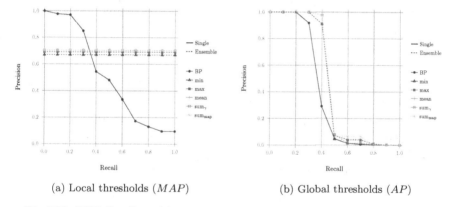

(a) Local thresholds (MAP) (b) Global thresholds (AP)

Fig. 7.35: BOT: Recall-precision curves using BP and BP-based ensemble methods.

tional overhead for combining the resulting distances, which is, however, negligibly small when compared to the cost of matching graphs). Note, however, that dissimilarities between two (or more) different graph representations can be computed independently of each other. Hence, the required computations can easily be parallelised.

7.7.2 *Ensemble Methods Based on HED, CED, and BP2*

Table 7.43 provides a summary of ensemble methods based on the four different graph matching algorithms BP, HED, CED, and BP2. Note that

Table 7.43: MAP of the KWS system using different ensemble methods in comparison with BP-based ensemble methods. With ± we indicate the relative percental gain or loss in the accuracy of the ensemble methods when compared with BP. The best result per column is shown in bold.

Method	GW		PAR		AK		BOT	
	MAP	±	MAP	±	MAP	±	MAP	±
BP	71.09		79.38		84.77		68.88	
HED	73.64	+2.55	80.58	+1.20	83.80	−0.97	71.00	+2.12
CED	**74.71**	+3.61	**81.08**	+1.70	**85.46**	+0.69	**72.51**	+3.63
BP2	72.39	+1.30	74.82	−4.56	78.73	−6.04	70.99	+2.11

we use the BP-based ensemble methods as a reference system. We observe that HED-based ensemble methods lead to an improvement in three out of four cases when compared with the BP-based ensemble. In the case of CED-based ensemble methods, we even observe an improvement in all cases of up to 4 basis points when compared with the BP-based ensemble. In contrast to this, in two cases, the BP2-based ensemble methods lead to a decline in accuracy of up to 6 basis points when compared with BP. Overall, we assume that ensemble-methods are especially beneficial with respect to documents with large intraword variations (style and size) and/or other variations (i.e. skewed scanning, imperfect segmentation, signs of degradation, etc.). That is, a single method is often largely affected by such variations, and thus, ensemble methods are able to reduce such negative effects. For more details with respect to the results obtained by the ensemble methods we refer to Chapter 11.

7.8   Cross-Evaluation of Graph-Based Systems

In this section, a cross-evaluation on all four manuscripts is conducted, as shown in Fig. 7.36. That is, the optimised parameter values for HED (see Table 7.44) are taken from one manuscript (e.g. GW) and employed on all three remaining manuscripts (e.g. PAR, AK, and BOT). Note that we limit this evaluation to local thresholds (i.e. MAP). By means of this evaluation, we aim to investigate the generalisability of the proposed graph-based KWS framework with respect to new and unseen manuscripts. In particular, we want to verify whether novel manuscripts can be retrieved without optimisation of the cost model.

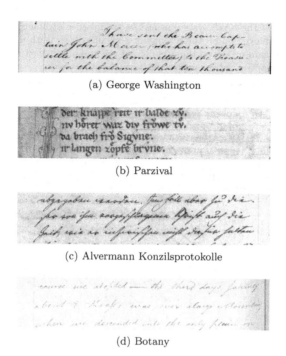

(a) George Washington

(b) Parzival

(c) Alvermann Konzilsprotokolle

(d) Botany

Fig. 7.36: Exemplary excerpts of the four historical manuscripts.

Table 7.44: Optimal cost function parameter for graph edit distance computation using HED.

Method	GW				PAR				AK				BOT			
	τ_v	τ_e	α	β	τ_v	τ_e	α	β	τ_v	τ_e	α	β	τ_v	τ_e	α	β
Keypoint	8	4	0.5	0.1	8	1	0.1	0.5	16	1	0.3	0.1	8	4	0.3	0.1

In Table 7.45 we present the results of the cross-evaluation for all manuscripts (columns) using cross-evaluated parameters (rows). In the case of GW and PAR, we observe a certain decline with respect to the MAP of the optimised parameter settings on the respective datasets, whereas the rate of accuracy in the case of AK and BOT remains nearly identical. The same effects can also be observed in the recall-precision curves in Fig. 7.37.

If we analyse the cross-evaluation results in greater detail, we observe a deterioration in the MAP of minus 8 to 0 basis points in the case of GW. In particular, parameter settings taken from BOT lead to a small

Table 7.45: *MAP* of the KWS system using the optimised cost function parameters of one manuscript employed on all three remaining manuscripts. With ± we indicate the absolute percental gain or loss in the accuracy of the cross-evaluated manuscript when compared with the optimised parameter settings (shown in bold).

Optimised on	GW *MAP*	±	PAR *MAP*	±	AK *MAP*	±	BOT *MAP*	±
GW	**69.28**		63.39	−5.84	79.54	−0.18	51.21	−0.48
PAR	64.84	−4.44	**69.23**		79.73	+0.01	51.12	−0.57
AK	61.45	−7.82	56.93	−12.30	**79.72**		50.81	−0.88
BOT	69.44	+0.17	62.40	−6.83	80.28	+0.56	**51.69**	

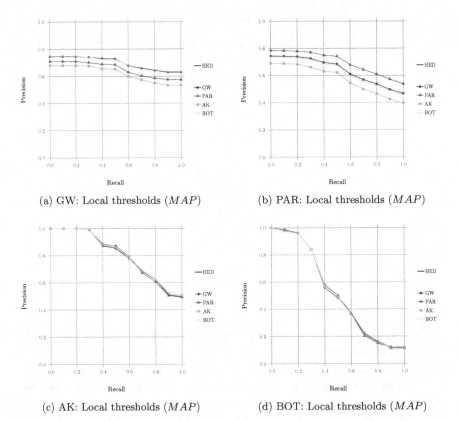

(a) GW: Local thresholds (*MAP*)   (b) PAR: Local thresholds (*MAP*)

(c) AK: Local thresholds (*MAP*)   (d) BOT: Local thresholds (*MAP*)

Fig. 7.37: Cross-evaluation: recall-precision curves using HED graph-matching algorithm.

improvement, whereas parameter settings taken from AK lead to a decrease of about 8 basis points. Possible reasons for this effect can be found in the optimal parameter settings, as shown in Table 7.44. In the case of GW and BOT, only the α-weighting factor is slightly different. In comparison to this, the cost model for AK clearly differs with respect to τ_v and τ_e when compared with the optimal cost model for GW. However, differences in the β-weighting factor also have an negative impact, as shown in the example of PAR-optimised parameter values.

We conclude that changes with respect to α (i.e. weighting factor between node and edge edit costs) seem to have a minor impact. In contrast to this, changes in the node and edge edit costs τ_v and τ_e, as well as β (i.e. weighting factor between x- and y-coordinates) seem to have a greater impact.

A similar effect can be observed in the case of PAR, where we observe a decrease in accuracy of 6 to 12 basis points. Likewise, with GW, the MAP mostly declines in the case of parameter settings taken from AK. That is, parameters optimised for PAR make use of $\tau_v = 8$ and $\beta = 0.5$, while parameters optimised for AK make use of $\tau_v = 16$ and $\beta = 0.1$. We assume that these larger node edit costs, in addition to the lower importance of the x-coordinates, lead to more substitutions between nodes of different letters (see also Fig. 6.3 and 6.4 for illustrations), and thus to an overall lower MAP. This effect is also supported by Fig. 7.36, where we see that the handwriting in PAR is more dense than in the remaining manuscripts. Hence, variations in the x-direction are more relevant (thus the higher β value), than in less dense handwriting such as in GW, AK, and BOT.

In the cases of AK and BOT, we observe nearly identical accuracy rates over all four employed parameter settings. Hence, both documents seem to be less sensitive with respect to variations in the cost model.

We conclude that the weighting parameter α has only a minor influence on the MAP in most cases. An example is given in the cases of GW and BOT, where similar accuracy rates can be observed with $\alpha = 0.5$ and $\alpha = 0.1$. In contrast to this, the weighting parameter β has a special influence on the density of the handwriting in particular, as shown in the example of PAR. That is, the x-direction becomes more important in dense handwriting, and thus higher parameter values for β should be chosen. Finally, the cost parameters τ_v and τ_e are crucially influenced by the size of the handwriting. An example is given in the case of AK, where the handwriting is characterised by flourish and thus node substitutions should be preferably be allowed, even for dissimilar nodes (i.e. we choose higher values for τ_v). In contrast to this, GW and especially PAR are not characterised so much by flourish and thus node substitutions should not be allowed to be made too easily (i.e. lower values for τ_v should be defined). Hence, if we relax the cost functions too much (i.e. employ settings optimised on AK and on GW for PAR) the MAP is negatively affected.

Certainly, the cost model for GED depends on the data, that is, on the characteristics of the handwriting in this case. We see that the weighting factor β should reflect the density of the handwriting, while cost parameters τ_v and τ_e should reflect the sizing and flourish of the handwriting. Hence, novel manuscripts could possibly be retrieved without exhaustive optimisation steps but rather by directly adapting the cost model to the characteristics of the handwriting.

7.9 Quantitative and Qualitative Summary of Graph-Based Systems

In this section we quantitatively and qualitatively summarise the results obtained by all the different graph-based KWS systems. With respect to the quantitative summary, we compare the best results per graph-matching algorithm with the best result of the baseline system BP. In particular, we compare improvements with respect to both KWS accuracy and runtime. In the second part, we qualitatively evaluate some retrieval results. That is, we visually compare the top ten retrieval results for different random keywords.

7.9.1 *Quantitative Summary*

In Table 7.46, results of all graph-based KWS systems evaluated in this book are summarised for local and global thresholds and compared with the baseline algorithm BP. Regarding the BP-Q method, we observe a decline in accuracy in six out of eight cases. Especially on the AK and BOT manuscripts we observe a decline of 4 basis points when compared with BP. In contrast to BP-Q, both fast rejection methods BP-FRN and BP-FRE lead to accuracy improvements in all threshold scenarios and on all manuscripts. With respect to the linear time graph matching algorithm PGD-Node, we have to consider a decline in accuracy in all cases of up to 18 basis points. These negative effects of PGD-Node can be compensated by means of PGD-Edge, where we observe an improvement in accuracy in the

Table 7.46: *MAP* and *AP* of different graph-based KWS methods. With ± we indicate the relative percental gain or loss in the accuracy of a particular method when compared with the baseline system BP. The best result per column is shown in bold (for single and ensemble systems).

Method	GW				PAR				AK				BOT			
	MAP	±	*AP*	±	*MAP*	±	*AP*	±	*MAP*	±	*AP*	±	*MAP*	±	*AP*	±
								Reference								
BP	66.08		54.37		66.23		64.23		77.24		76.32		49.57		38.33	
								Single								
BP-Q	65.92	−0.16	54.43	+0.06	64.62	−1.61	67.82	+3.59	75.03	−2.21	72.40	−3.92	47.24	−2.33	35.39	−2.94
BP-FRN	69.81	+3.73	56.45	+2.08	71.09	+4.86	71.77	+7.54	81.51	+4.27	79.01	+2.69	56.10	+6.53	39.85	+1.52
BP-FRE	70.61	+4.52	57.04	+2.67	72.03	+5.80	71.48	+7.25	81.51	+4.27	79.91	+3.59	**57.14**	+7.57	40.48	+2.15
PGD-Node	58.47	−7.62	47.13	−7.24	50.05	−16.18	45.91	−18.32	72.36	−4.88	69.52	−6.80	46.13	−3.44	35.48	−2.85
PGD-Edge	68.78	+2.70	58.10	+3.73	62.62	−3.61	71.23	+7.00	76.98	−0.26	74.76	−1.56	52.11	+2.54	**41.21**	+2.88
HED	69.28	+3.19	57.38	+3.01	72.82	+6.59	76.69	+12.46	81.06	+3.82	79.81	+3.49	51.74	+2.17	40.29	+1.96
CED	**71.23**	+5.15	**59.15**	+4.78	**74.10**	+7.87	77.94	+13.71	**82.69**	+5.45	**81.46**	+5.14	52.45	+2.88	40.34	+2.01
BP2	68.42	+2.33	52.47	−1.90	68.69	+2.46	69.54	+5.31	75.46	−1.78	73.85	−2.47	50.94	+1.37	39.25	+0.92
								Ensemble								
BP	71.09	+5.01	59.15	+4.78	79.38	+13.15	85.43	+21.20	84.77	+7.53	**83.34**	+7.02	68.88	+19.31	48.29	+9.96
HED	73.64	+7.56	63.84	+9.47	80.58	+14.35	80.58	+16.35	83.80	+6.56	80.97	+4.65	71.00	+21.43	49.88	+11.55
CED	**74.71**	+8.62	**65.21**	+10.84	**81.08**	+14.85	**86.94**	+22.71	**85.46**	+8.22	83.29	+6.97	**72.51**	+22.94	50.33	+12.00
BP2	72.39	+6.31	63.32	+8.95	74.82	+8.59	81.37	+17.14	78.73	+1.49	78.84	+2.52	70.99	+21.42	**50.79**	+12.46

majority of the cases. In the case of HED, KWS accuracy is substantially improved in all threshold scenarios and on all manuscripts. We assume that this is mostly due to the property of multiple node assignments allowed by HED, and thus, variations caused by scaling and style can be compensated well. Moreover, accuracy can be further improved by up to 14 basis points by means of CED when compared to BP. That is, the consideration of larger node contexts generally allows us to improve KWS accuracy. Finally, we observe quite a similar rate of accuracy of BP2 when compared with the baseline algorithm BP. In summary, CED achieves the overall best accuracy rates in six of eight cases and can be seen as the clear winner among all systems.

Regarding the results of the ensemble methods, we observe even larger accuracy improvements in all threshold scenarios and on all manuscripts. In particular, we observe an improvement in all ensemble methods when compared to the respective single methods. Especially, the CED-based ensemble methods lead to the overall best KWS accuracy rate in six out of eight cases. Using an ensemble on the BOT manuscript, we observe improvements of up to 23 basis points when compared to the corresponding single systems. Hence, we conclude that ensemble methods have a beneficial impact for manuscripts with rather wide variations in style and scaling, and large signs of degradation.

In a second evaluation, we compare the different graph matching algorithms with respect to their runtime, as shown in Table 7.47. That is, we compare the systems with respect to the lower (i.e. minima, denoted by *lower speed-Up factor* (SF_L)) and upper (i.e. maximal, denoted by *upper speed-Up factor* (SF_U)) speed-up factors when compared with the cubic time baseline system BP. The different graph matching algorithms are sorted by means of their computational complexity. Regarding the cubic time algorithm BP-Q, we observe speed-up factors of 4 to 22 when compared with the baseline system BP. Quite similar speed-up factors can be obtained for BP-FRN and BP-FRE. We observe substantially smaller runtimes with the quadratic time algorithms HED and BP2. That is, with these procedures we observe speed-up factors of between 55 and 279 when compared with BP. Slightly smaller speed-up factors are achieved by CED, where the larger context leads to more computations of node assignments and therefore also to lower runtimes. Finally, we observe very high speed-ups with the novel linear time graph dissimilarities PGD-Node and PGD-Edge. That is, these approaches are up 5,000 times faster than the cubic time algorithm BP.

Table 7.47: SF_L and SF_U of different graph-based KWS methods sorted by their computational complexity.

Method	GW		PAR		AK		BOT	
	SF_L	SF_U	SF_L	SF_U	SF_L	SF_U	SF_L	SF_U
Cubic time complexity								
BP								
BP-Q	15	17	21	22	10	16	4	6
BP-FRN	3	14	4	16	7	11	6	7
BP-FRE	7	24	6	14	9	13	7	7
Quadratic time complexity								
HED	88	116	79	112	55	91	91	230
CED	18	53	14	57	8	44	16	118
BP2	91	122	78	118	56	92	110	279
Linear time complexity								
PGD-Node	1,058	2,011	896	2,487	418	1,062	965	4,051
PGD-Edge	767	1,077	868	1,302	310	951	1,065	5,260

To sum up this section, with respect to the baseline framework BP, we observe several improvements in various extensions and alternatives to BP. First, we observe improvements in accuracy by most graph dissimilarities. HED-based approaches, for instance, lead to substantially higher accuracy rates than BP due to their more flexible node assignment approach. Moreover, CED, which considers a larger node context, allows us to clearly increase the accuracy rates in both local and global threshold scenarios. We also observe the beneficial effects of the proposed ensemble methods when compared with the individual methods. That is, ensemble methods allow clear improvements in accuracy, especially in manuscripts with large intraword variabilities like BOT. Moreover, we observe clear improvements with respect to the runtime. In particular, the novel linear time graph dissimilarities PGD-Node and PGD-Edge allow speed-up factors of more than 5,000 when compared with the cubic time algorithm BP. However, the quadratic time algorithms HED, CED, and BP2 also allow clear speed-ups of the KWS system.

7.9.2 *Qualitative Summary*

In this section, we provide a qualitative evaluation of our graph-based KWS framework on all manuscripts, namely GW, PAR, AK, and BOT. To this end, we evaluate the top ten results qualitatively (i.e. visually) for five random keywords (i.e. queries) per manuscript. We make use of the BP graph matching algorithm as a basic dissimilarity model. Note, however, that similar, probably even better, observations can be made for all of the proposed graph-based KWS approaches.

In Fig. 7.38, we provide a qualitative evaluation for the GW manuscript. We observe that the first and the second retrieved keyword (*Instructions.*) correspond to the actual query word. Note that in the third and fourth

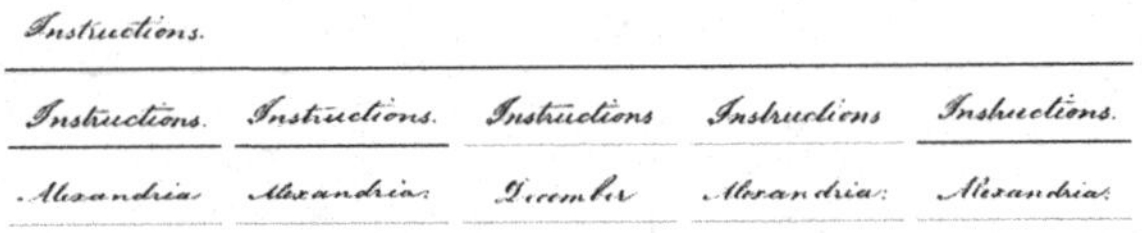

(a) Query: *Instructions.*

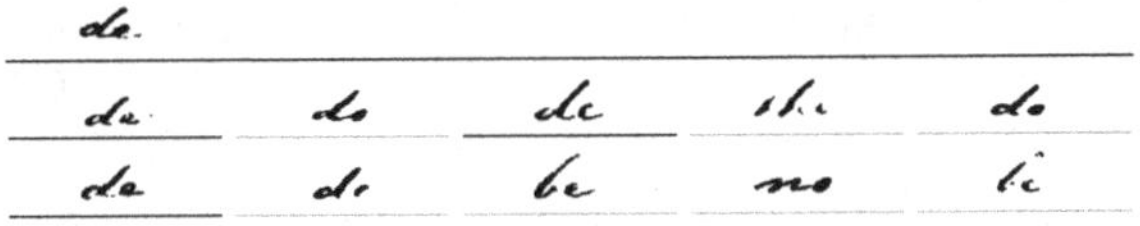

(b) Query: *de*

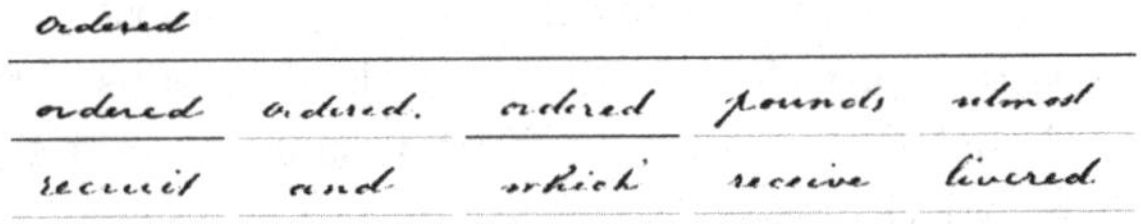

(c) Query: *ordered*

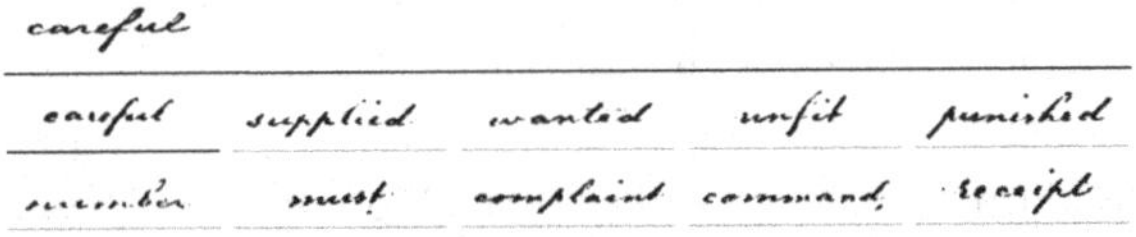

(d) Query: *careful*

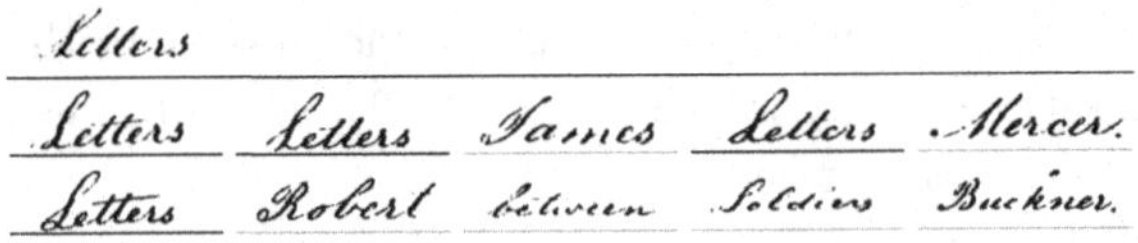

(e) Query: *Letters*

Fig. 7.38: GW: Top ten results for given keywords using the BP graph-matching algorithm. True positives are underlined in black and false positives are underlined in grey.

retrievals we observe false positives due to the missing punctuation at the end of the word (see also the query *ordered* in Fig. 7.38c for a similar observation). Overall, we observe that all ten retrieved words are visually very similar to the query word. Similar effects of visually similar false positives can be observed in the remaining cases. Note especially the query word *de* where the differences to the false positives (e.g. *do*, *the*, *be*, and *no*) are often hardly visible to humans.

In the PAR manuscript, we observe quite similar effects even though the intraword variation is rather small, as shown in Fig. 7.39. With respect

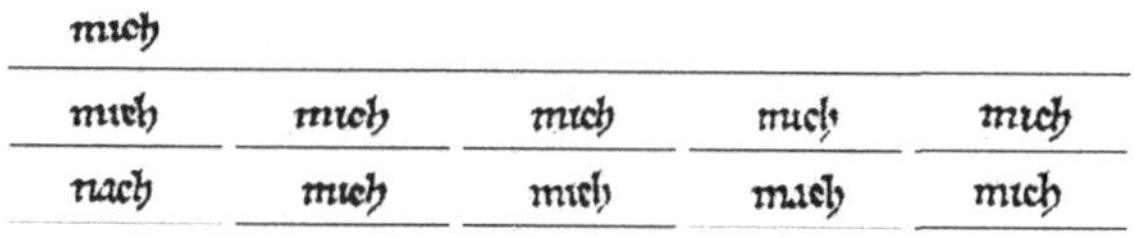

(a) Query: *mich*

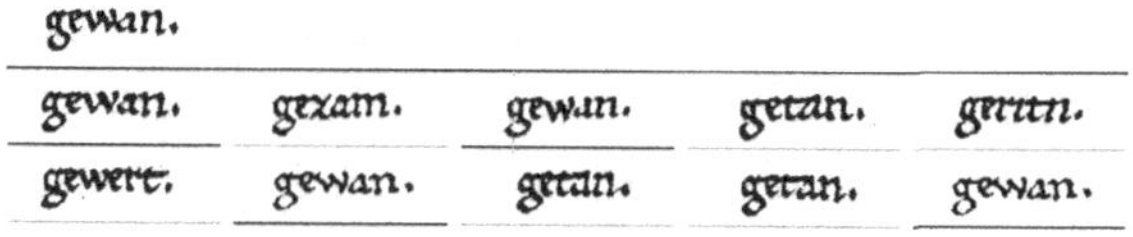

(b) Query: *gewan.*

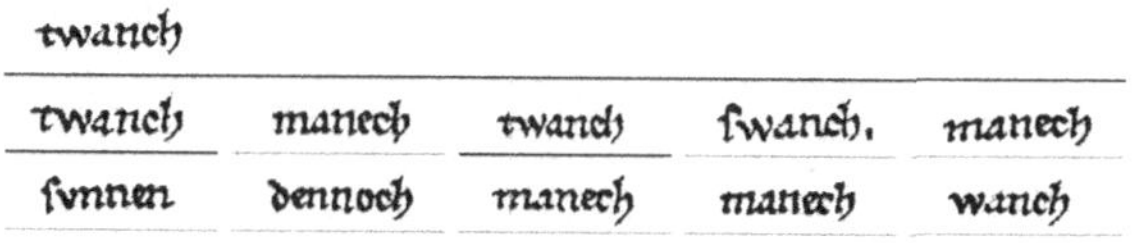

(c) Query: *twanch*

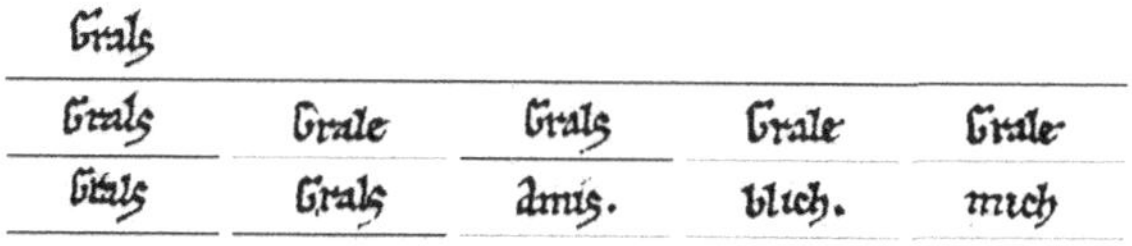

(d) Query: *Grals*

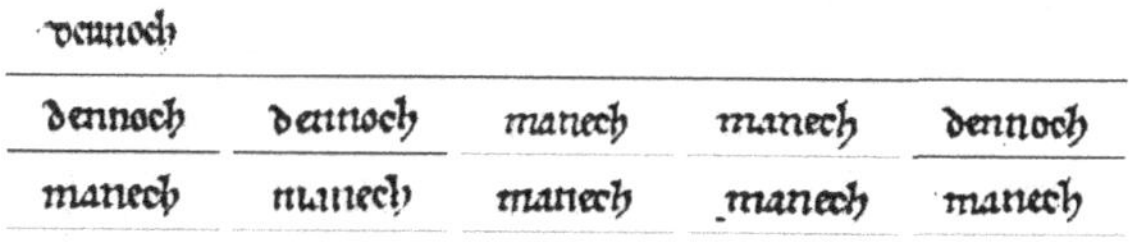

(e) Query: *dennoch*

Fig. 7.39: PAR: Top ten results for given keywords using the BP graph-matching algorithm. True positives are underlined in black and false positives are underlined in grey.

to false positives, we observe that differences between a query word and false positives are often due to a single character, such as, for example, in the case of the query *Grals* and the first false positive *Grale*.

The visual results on the AK manuscript are shown in Fig. 7.40. Note the additional challenge placed on this dataset due to imperfect word segmentation (in the case of the query *Warneckros*, for instance). In particular, we observe fragments of ink (i.e. underlining or other word parts) that

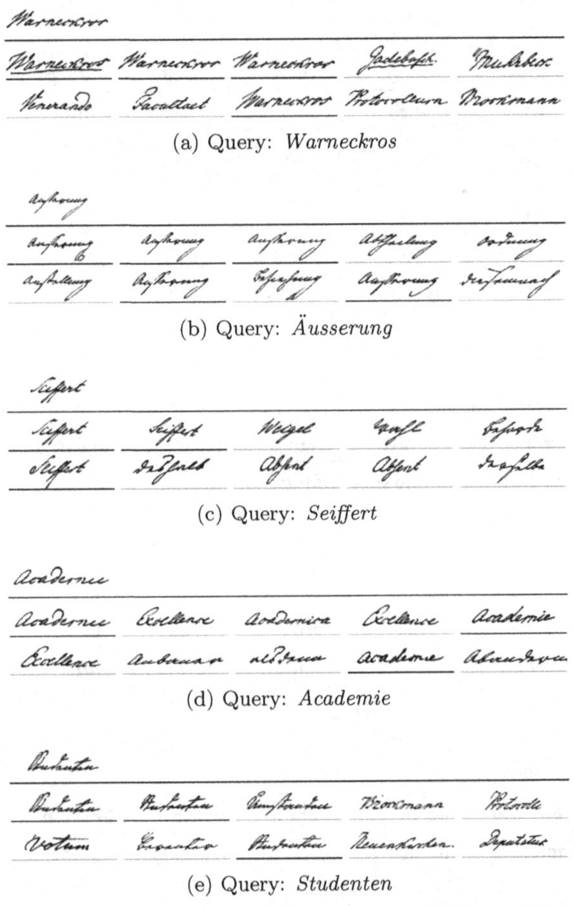

(a) Query: *Warneckros*

(b) Query: *Äusserung*

(c) Query: *Seiffert*

(d) Query: *Academie*

(e) Query: *Studenten*

Fig. 7.40: AK: Top ten results for given keywords using the BP graph-matching algorithm. True positives are underlined in black and false positives are underlined in grey.

do not belong to the actual word in the case of the first, fourth, and fifth word of the retrieval results. Moreover, we can observe strong variations in the writing style in the case of the query *Academie*. That is, the letter 'd' is written quite differently in the query word and in the second and third true positive.

In Fig. 7.41, we provide selected qualitative results on the BOT manuscript. We observe, for instance, that the query *William* retrieves all three instances as top 3 results, while several false positives can be

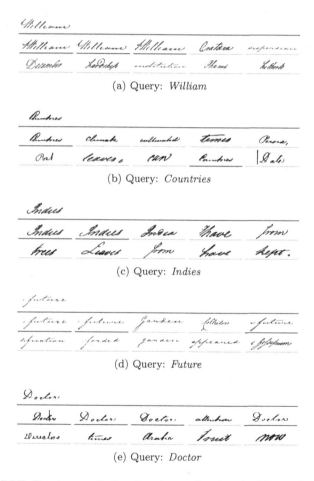

(a) Query: *William*

(b) Query: *Countries*

(c) Query: *Indies*

(d) Query: *Future*

(e) Query: *Doctor*

Fig. 7.41: BOT: Top ten results for given keywords using the BP graph-matching algorithm. True positives are underlined in black and false positives are underlined in grey.

observed in the case of the query *Countries*. In this particular case, we observe that all the false positives are actually quite different to each other. That is, in most cases we observe that false positives are based on rather homogenous groups of words. An example is given with the query word *Indies* where the false positives *have* and *from* are represented twice in the list of top ten results. Similar to the AK manuscript, we observe additional variations caused by imperfect word segmentations on this dataset.

Overall, we conclude that all four manuscripts are somewhat negatively affected by false positives to a certain degree. That is, the proposed approach of graph-based KWS is somewhat affected by topologically similar words. Moreover, imperfect word segmentations and signs of degradation also lead to false positives. In general, this effect could be reduced by providing more query words for the same keyword such that the query words would cover different writing styles and other word variations.

Note that the false positives are also affected by the chosen cost model for GED. In general, low substitution costs for nodes and edges allow higher variations, and thus, also more false positives. Vice versa, defining a cost model with high costs is too rigid as we observe certain intraword variations in our application that need to be captured by means of a more flexible cost model.

In Fig. 7.42a and 7.42b, we present retrieval results on the validation set with a suboptimal cost model ($\tau_v = 1$, $\tau_e = 1$) and an optimal cost

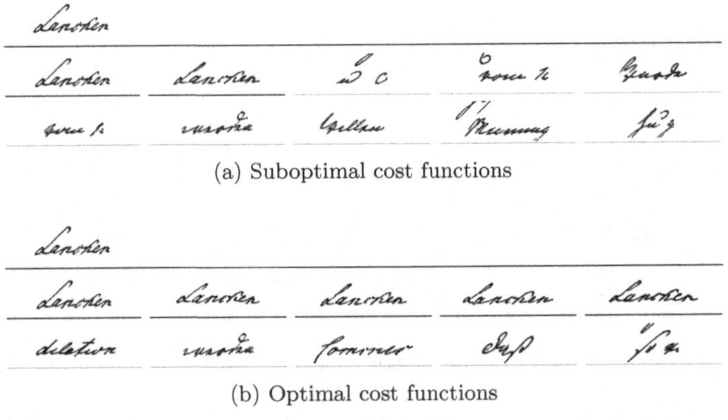

(a) Suboptimal cost functions

(b) Optimal cost functions

Fig. 7.42: Query: *Lancken* with (a) suboptimal and (b) optimal cost function parameters. True positives are underlined in black and false positives are underlined in grey.

model ($\tau_v = 16$, $\tau_e = 16$), respectively. Note that the low insertion and deletion costs of the suboptimal cost model leads to few substitutions only, and thus, we observe only two true positives in this case. In contrast to this, we observe that higher insertion and deletion costs lead to more substitutions and better results in general. As a result, we observe not only more true positives (five rather than two) but also very visually similar false positives. Hence, a tradeoff between a flexible cost model and a false positive rate generally needs to be found. To this end, we have thoroughly validated our cost model (see Section 7.2.1).

7.10 Comparison of Our Framework with Reference Systems

To conclude this chapter, we provide an experimental comparison of the proposed graph-based systems with different state-of-the-art reference systems. First, we compare the graph-based approaches with template-based reference systems for the GW and PAR manuscripts. Second, a comparison with state-of-the-art learning-based approaches is given for the AK and BOT manuscripts.

7.10.1 *Reference Systems*

We employ two types of reference systems, namely four template-based systems using *dynamic time warping* (DTW) [51–53, 56], and three learning-based KWS systems using *support vector machines* (SVMs) and *convolutional neural networks* (CNNs) [65, 66, 69]. In general, learning-based KWS leads to higher accuracy rates than template-based systems. However, this inherent advantage is accompanied by a loss of generalisability and flexibility due to the need for labelled training data. For a full comparison and further details regarding template- and learning-based KWS, we refer to Chapter 2.

Note that the results of the template-based reference systems are available for GW and PAR only[5], while the results of all learning-based systems are available for AK and BOT only. In the following two subsections both types of reference systems are briefly reviewed.

---

[5] For an unbiased comparison we consider systems that make use of unlabelled images only (i.e. template-based). That is, we do not consider some recent learning-based approaches tested on GW and PAR [32, 56, 123].

7.10.1.1 *Template-based Reference Systems*

With respect to template-based reference systems, we consider the most popular matching algorithm, namely *dynamic time warping* (DTW). In particular, DTW optimally aligns (warps) two sequences of features vectors $X = \{\mathbf{x}_1, \ldots, \mathbf{x}_m\}$ and $Y = \{\mathbf{y}_1, \ldots, \mathbf{y}_n\}$ along one common time axis using a dynamic programming approach. For our reference systems these feature vectors either consist of nine different geometrical features [51] (reference system denoted by DTW'01), *histograms of oriented gradients* (HoG) features [52, 53] (denoted by DTW'08 and DTW'09), or deep learning features [56] (denoted by DTW'16). For all systems, the alignment cost $d(\mathbf{x}, \mathbf{y})$ between each vector pair $(\mathbf{x}, \mathbf{y}) \in \mathbb{R}^m \times \mathbb{R}^n$ is given by the squared Euclidean distance. The DTW distance $D(X, Y)$ between two sequences of feature vectors is then given by the minimum alignment cost found by dynamic programming. Formally,

$$D(X, Y) = \sum_{k=1}^{K} d(\mathbf{x}_{i_k}, \mathbf{y}_{j_k})$$

where K is the length of the optimal warping path $((i_1, j_1), \ldots, (i_K, j_K))$ [46]. A *Sakoe-Chiba band* [300] that constrains the warping path is often applied to speed up this procedure. For further details regarding these template-based reference systems we refer to Section 2.4.1.

7.10.1.2 *Learning-based Reference Systems*

With respect to learning-based reference systems, we take into account three state-of-the-art learning-based methods, namely CVCDAG [69], PRG [65], and QTOB [66]. CVCDAG is based on the *pyramid histogram of characters* (PHOC) features used in conjunction with a *support vector machine* (SVM) [69]. PHOC is a word string embedding approach based on five different splitting levels, i.e. on level n a word image is split into n subparts for which a histogram for the number of character occurrences is created. In PRG, the same features are used for training in a *convolutional neural network* (CNN), the so-called PHOCNet [65]. Another CNN is used in QTOB by means of a triplet network approach [66]. For further details about these learning-based reference systems we refer to Section 2.4.2[6].

---

[6]Global threshold results are not available for AK and BOT as the ICFHR2016 competition is based on local thresholds only.

7.10.2 *Comparison with Template-Based Reference Systems*

We compare all graph-based approaches including all ensemble methods with the four template-based reference systems for the GW and PAR manuscripts, as shown in Table 7.48. We observe that all single graph systems (except PGD-Node) outperform the first two reference systems DTW'01 and DTW'08 on both manuscripts. Moreover, with respect to the third and fourth reference systems DTW'09 and DTW'16, we observe that at least one graph-based approach (six systems in total) outperforms these template-based approaches. This is particularly interesting as DTW'16 performs an unsupervised feature learning step using unlabelled data, while our graph-based methods are based on spatial information only.

We have already seen that the accuracy rates achieved by the different graph-based ensemble methods are generally higher than the individual

Table 7.48: *MAP* for graph-based KWS systems in comparison with four template-based reference systems. The first, second, and third best systems (single and ensemble-based) are indicated by (1), (2), and (3), respectively.

Method	GW		PAR		Average	
Reference (template)						
DTW'01	45.26		46.78		46.02	
DTW'08	63.39		47.52		55.46	
DTW'09	64.80		73.49	(2)	69.15	
DTW'16	68.64		72.38	(3)	70.51	
Graph (single)						
BP	66.08		66.23		66.16	
BP-Q	65.92		64.62		65.27	
BP-FRN	69.81	(3)	71.09		70.45	
BP-FRE	70.61	(2)	72.03		71.32	(2)
PGD-Node	58.47		50.05		54.26	
PGD-Edge	68.78		62.62		65.70	
HED	69.28		72.82		71.05	(3)
CED	71.23	(1)	74.10	(1)	72.67	(1)
BP2	68.42		68.69		68.55	
Graph (ensemble)						
BP	71.09		79.38	(3)	75.24	(3)
HED	73.64	(2)	80.58	(2)	77.11	(2)
CED	74.71	(1)	81.08	(1)	77.89	(1)
BP2	72.39	(3)	74.82		73.61	

systems. Due to these advantages, we observe that all ensemble methods (regarding the basic graph matching algorithm) outperform all four reference systems on both manuscripts. With respect to CED-based ensemble methods, we observe the best KWS accuracy overall on both manuscripts. We can report improvements of about 6 and 7 basis points when compared to the best DTW-based reference systems. Note that the first, second, and third best single and ensemble-based systems are indicated by (1), (2), and (3), respectively.

In Table 7.49, we also provide an empirical runtime evaluation using `Keypoint` graphs on the GW manuscript. To this end, we measure the average matching time for 10,000 pairs of feature vectors (in the case of DTW systems) or graphs (in the case of our framework). We observe that the cubic time algorithm BP (and all related methods) lead to clearly higher matching times than the DTW-based reference system. However, with respect to the quadratic graph-matching approaches HED and BP2, we observe the faster matching times of our approach in three out of four cases. Currently, only DTW'01 is faster than our system, which is based on rather low-dimensional feature vectors (i.e. nine geometrical characteristics). It seems that the number of nodes in our specific graphs is often smaller than the number of window positions in DTW-based systems, and thus,

Table 7.49: Matching times (ms) for graph-based KWS systems in comparison with four template-based reference systems.

Method	Matching time (ms)
Reference (template)	
DTW'01	0.7
DTW'08	5.6
DTW'09	10.2
DTW'16	7.7
Graph	
BP	303.0
BP-Q	17.7
BP-FRN	39.5
BP-FRE	14.2
PGD-Node	0.3
PGD-Edge	0.4
HED	3.2
CED	7.5
BP2	3.3

quadratic time graph-matching approaches are faster than a DTW-based reference system with the same time complexity.

Moreover, we observe very low matching times with respect to PGD-Node and PGD-Edge (both offer linear time complexity). These runtimes are even lower than those of the fastest reference system DTW'01.

We conclude that graph-based approaches lead to higher KWS accuracy rates than DTW systems in general. Moreover, by means of special approximation algorithms applied to (smaller) graphs, we observe lower runtimes when compared with state-of-the-art DTW approaches.

7.10.3 *Comparison with Learning-Based Reference Systems*

Finally, the novel graph-based KWS systems are compared with three state-of-the-art learning-based reference systems for AK and BOT manuscripts, as shown in Table 7.50. Regarding the single graph-based approaches, we observe the clear advantage of learning-based systems on the BOT manuscript. While learning-based approaches achieve KWS accuracies between 55% and 90%, our graph-based methods achieve a maximum accuracy rate of about 57%. We conclude that learning-based approaches in particular can better handle the large intraword variations on this manuscript (in style and size). Nevertheless, we observe that both fast rejection approaches BP-FRN and BP-FRE achieve better results than the learning-based reference system QTOB. We observe a substantially smaller difference between the graph-based and learning-based approaches on the AK manuscript. That is, on these datasets the learning-based approaches lead to accuracy rates of between 78% to 96%, while our single graph-based approaches achieve accuracy rates of between 75% and 83%. In particular, we observe that the learning-based systems CVCDAG and QTOB can be outperformed by graph-based approaches in four and three cases, respectively.

A substantial increase in performance can be observed on both AK and BOT with our novel graph-based KWS ensemble. We observe accuracy rates of between 79% and 85% on the AK manuscript, and of between 69% and 73% on the BOT manuscript. In particular, all graph-based ensemble methods achieve better results than the learning-based framework QTOB for both manuscripts. Moreover, we observe that all graph-based ensemble methods outperform CVCDAG on the AK manuscript. This is quite astonishing as our proposed methods — in contrast to the learning-based approaches — are not trained on labelled data. In particular, the

Table 7.50: *MAP* for graph-based KWS systems in comparison with three state-of-the-art learning-based reference systems. The first, second, and third best systems are indicated by (1), (2), and (3), respectively.

Method	AK		BOT		Average	
			Reference (learning)			
CVCDAG	77.91		75.77	(2)	76.84	
PRG	96.05	(1)	89.69	(1)	92.87	(1)
QTOB	82.15		54.95		68.55	
			Graph (single)			
BP	77.24		49.57		63.41	
BP-Q	75.03		47.24		61.14	
BP-FRN	81.51		56.10		68.81	
BP-FRE	81.51		57.14		69.33	
PGD-Node	75.46		46.13		60.80	
PGD-Edge	76.98		52.11		64.55	
HED	81.06		51.74		66.40	
CED	82.69		52.45		67.57	
BP2	75.46		50.94		63.20	
			Graph (ensemble)			
BP	84.77	(3)	68.88		76.83	
HED	83.80		71.00		77.40	(3)
CED	85.46	(2)	72.51	(3)	78.99	(2)
BP2	78.73		70.99		74.86	

accuracy of learning-based approaches is often dependent on the size of the labelled training data[7]. However, in the case of handwritten historical documents, the acquisition of large sets of labelled training data is often a labour-intensive and costly procedure. Moreover, the size of the training data is often inherently limited by the size of the document. Hence, our learning-free approach offers clear advantages in this particular setting.

7.11 Summary

The experimental evaluation in this chapter is divided into two parts. First, we compare different graph-based approaches with each other. In particular, we compare the accuracies for different graph-matching

---

[7]In the case of PRG, smaller accuracy rates can be clearly observed in the case of penalised scores that take the size of the training data into consideration, see [80].

procedures for a local and global threshold scenario, as well as speed-up factors when compared to a graph-based baseline system. Second, we compare our novel graph-based framework with four state-of-the-art template-based approaches on the *George Washington* (GW) and *Parzival* (PAR) manuscripts, and we provide a comparison of our methods with recent learning-based approaches for the *Alvermann Konzilsprotokolle* (AK) and *Botany* (BOT) manuscripts.

In the first experimental evaluation we compare four different graph representations (i.e. Keypoint, Grid, Projection, and Split) with each other in a *keyword spotting* (KWS) experiment using the cubic time graph-matching algorithm *bipartite graph edit distance* (BP). We observe that either Keypoint or Projection graphs lead to the highest accuracies in both threshold scenarios. However, Split graphs achieve comparable accuracies. For Grid graphs we observe certain deficiencies, especially on the PAR manuscript.

In the empirical evaluation, different speed-up approaches are examined and compared. A first approach aims to speed up the matching procedure by means of a quadtree segmentation approach (denoted by BP-Q). In this procedure, graph matchings are conducted on smaller subgraphs rather than on complete graphs. For all manuscripts we observe speed-up factors between 4 and 22 when compared with the baseline system BP. However, this speed-up is accompanied by a minor decline in KWS accuracy overall. A second type of speed-up approaches is based on fast rejections. To this end, graphs are first filtered by means of a novel linear time graph dissimilarity *polar graph dissimilarity* (PGD) (denoted by PGD-Node and PGD-Edge). In particular, pairs of graphs are only matched if the node or edge distribution in a polar coordinate system are similar enough. By means of these filtering approaches (denoted by BP-FRN and BP-FRE), we observe speed-up factors between 3 and 24 when compared with the basic framework BP. Moreover, with this extension, we observe improvements in KWS accuracy of up to 8 basis points for both threshold scenarios. We also employ the PGD-Node and PGD-Edge as isolated graph measures (rather than filters). We observe speed-up factors of up to 5,000 when compared with BP. However, this huge improvement with respect to runtime is accompanied with a clear accuracy decline in the case of PGD-Node. In contrast to this, we observe similar accuracy rates using PGD-Edge when compared with BP. That is, PGD-Edge allows both a massive speed-up of the KWS process and similar KWS accuracy rates as the baseline system BP.

Rather than speeding up the graph-based KWS system by means of different heuristics and filtering approaches, we evaluate different graph-matching algorithms with quadratic rather than cubic time complexity in our framework, namely *Hausdorff edit distance* (HED), *context-aware Hausdorff edit distance* (CED), and *bipartite graph edit distance 2* (BP2). With the HED method we observe both a clear improvement in accuracy and speed-up factors of up to 230 when compared with BP. In particular, HED allows multiple node assignments during the matching process, making it more robust towards scaling and intraword variations. Even larger improvements in KWS accuracy can be observed in the case of CED, where the node context — rather than adjacent edges only — is taken into consideration during the matching procedure. This larger node context allows us to reduce the approximation error with respect to the exact *graph edit distance* (GED), and thus improves KWS accuracy. Finally, with BP2 we observe very similar accuracy rates as for BP. However, due to the quadratic time complexity of BP2, we observe similar speed-up factors as with HED.

We also evaluate different ensemble methods that combine different distances obtained on different graph representations. The distances are aggregated by means of different statistical functions. We observe clear accuracy improvements with respect to all graph matching algorithms and manuscripts of up to 23 basis points in both threshold scenarios. In particular, we observe that the ensemble methods mean, sum_γ and sum_{map} are beneficial in most of the cases. Note, however, that ensemble methods come with an increased computational complexity due to several matchings of the involved graph representations (this drawback can, however, be compensated by means of parallel computations).

Moreover, the generalisability of the proposed KWS framework is evaluated empirically in a cross-evaluation. That is, three manuscripts are tested with parameters optimised on one specific manuscript. We observe that the employed cost model depends directly on the underlying characteristics of the handwriting. Hence, we conclude that novel manuscripts can be retrieved directly without exhaustive optimisation of the cost model.

We also compare our graph-based approaches with four *dynamic time warping* (DTW)-based reference systems on the GW and PAR manuscripts. We observe the clear advantages of most graph-based approaches. In particular, the graph-matching algorithms HED and CED, but also BP-FRN and BP-FRE outperform the statistical template-based approaches. With CED we observe accuracy rates of 71% and 74% on GW and PAR, respectively, while the best performing DTW-based approaches achieve accuracy rates of 69% and 73%. This is particularly impressive as the reference system makes use of an unsupervised deep learning approach for feature extraction. The advantages of our graph-based approaches are even more evident in the ensemble scenario. That is, the CED-based ensemble method leads to KWS accuracies of 75% and 81% on GW and PAR, respectively.

Our novel graph-based approaches can also keep up with the DTW systems in terms of runtime. In particular, the graph-matching algorithms with quadratic time, such as, for example, HED, CED, and BP2, are able to keep up with most DTW-based approaches. Moreover, the linear time algorithms PGD-Node and PGD-Edge lead to the lowest matching times of all examined approaches.

Finally, we compare the proposed graph-based approaches with different state-of-the-art learning-based approaches for KWS. We observe the high accuracy rates of the *convolutional neural network* (CNN)-based approach PRG of 96% and 90% on the AK and BOT manuscripts, respectively. Our best performing single graph-based approaches achieve accuracies of up to 82% and 57% on the same dataset. That is, we observe the advantages of the learning-based approach over the proposed graph-based methods, especially on the challenging BOT manuscript. However, by means of our graph-based ensemble methods we are able to substantially improve the accuracy on BOT. That is, we observe accuracy rates for the CED-based ensemble methods of 86% and 73% on AK and BOT, respectively. This system refers to the overall second best procedure. That is, our graph-based system turns out to be more accurate than advanced methods based on a *support vector machine* (SVM) (i.e. CVCDAG) or CNN (e.g. QTOB). This is an interesting observation, especially as our approach is not dependent on labelled training data. That is, the accuracy of learning-based approaches often crucially depends on the size of the training data available. However, the acquisition of training data in handwritten ancient documents is a labour-intensive and costly procedure. Moreover, the size of the training data is often inherently limited by the size of the document.

We conclude that graph-based approaches are able to outperform state-of-the-art template-based approaches with respect to both time and accuracy. Moreover, graph-based approaches are also able to keep up with, or even outperform, many of the learning-based reference systems. This makes graph-based KWS a very flexible and veritable alternative, especially as high accuracy rates can be achieved on manuscripts with wide variations without the need of an *a priori* learning step.

Chapter 8

Conclusion & Future Work

8.1 Conclusion

In recent decades, many handwritten historical documents have been digitised and made publicly available. However, the accessibility of these documents is often limited, and thus, we observe a certain gap between availability and accessibility. A reason for this might be the rather wide variations (in style, size, and signs of degradation) in the case of handwritten ancient documents, often making an automatic full transcription infeasible.

To bridge this gap, *keyword spotting* (KWS) is proposed as a flexible and more error-tolerant alternative to full transcriptions. KWS allows us to retrieve arbitrary keywords in a given document. Basically, most of the KWS approaches available to date are based on a statistical representation of certain characteristics of the handwriting. In contrast to this, we observe only few and rather limited approaches that make use of a structural representation of the document's content by means of strings, trees, or graphs. This is rather surprising as graphs offer some inherent advantages when compared with feature vectors. In particular, graphs are able to adapt their size and structure to the complexity of the underlying pattern. Moreover, graphs are able to represent binary relationships among subparts of the pattern. Both of these characteristics are highly beneficial for representing handwriting images. As a result, we see great potential for graph-based representations of handwriting and especially graph-based KWS. This book tries to close this research gap by developing and researching various novel graph-based approaches for KWS.

The proposed graph-based KWS framework is based on three subsequent process steps, namely image preprocessing, graph extraction from word images, graph matching, and querying. That is, document images are first preprocessed in order to minimise variations caused, for instance, by noisy background, skewed scanning, or signs of degradation. In our particular case, document images are first filtered by means of *difference of Gaussians* (DoG) filtering. Next, document images are binarised by means of global thresholding. Based on preprocessed document images, single word images are extracted by means of projection profiles. Finally, skew (i.e. the inclination of the document) is normalised by means of the linear regression of the lower baseline of a line of text.

In the second step (graph extraction), we employ four different graph representations to represent the inherent topological characteristics of handwritten word images by means of graphs. The first graph representation (denoted by `Keypoint`), is based on certain characteristics points (so-called keypoints) of the handwriting. That is, nodes represent these keypoints, while edges are inserted between interconnected keypoints. The second graph representation (denoted by `Grid`), is based on a grid-wise segmentation of handwriting images. For this formalism, nodes represent the centre of the mass of a segment, while edges are inserted between neighbouring segments and then reduced by means of a minimal spanning tree algorithm (actually transforming the graphs into trees). The two remaining formalisms (denoted by `Projection` and `Split`) are based on an adaptive segmentation by means of vertical and horizontal projection profiles. Nodes are then used to represent the centre of mass of a segment, while edges are inserted between nodes that are connected by a chain of foreground pixels.

For the creation of a retrieval index (i.e. an index sorted by the similarity of document words given a certain query word), we make use of inexact graph matching. Handwriting graphs are affected by certain variations (in style, scaling, etc.), making exact graph matching not feasible. We make use of *graph edit distance* (GED), a flexible and powerful paradigm for inexact graph matching [263]. Basically, GED measures the minimum amount of distortion needed to transform one graph into another, given a set of edit operations. However, exact GED is exponential with respect to the number of nodes under consideration. To overcome this limitation, a number of fast suboptimal algorithms for GED with cubic or quadratic time complexity have been proposed in recent years. These fast but suboptimal algorithms allow us to also employ GED in the domain of KWS, where rather large graphs are used to represent word images. First, we match a query graph with the set of document graphs by means of various algorithms. The resulting set of graph dissimilarities is then used as a retrieval index. In the best possible case this retrieval index consists of all instances of a keyword as its top n results.

The baseline KWS framework is based on the *bipartite graph edit distance* (BP) matching algorithm with cubic time complexity with respect to the number of involved nodes [93]. BP reduces the *quadratic assignment problem* (QAP) of GED computation to an instance of a *linear sum assignment problem* (LSAP) on the nodes and adjacent edges of the involved graphs. LSAPs — in contrast with QAPs — can be quite efficiently solved in cubic time by, for example, the Hungarian method [273]. Even though BP is quite fast, its cubic time complexity can still be a limiting factor in the case of large documents (with large amounts of words) or large graphs. In this book, two different speed-up approaches are proposed for BP. First, we employ a quadtree graph segmentation. Hence, graph matchings can be carried out on several smaller subgraphs instead of on one complete (typically large) graph. We denote this procedure by BP-Q. Second, we employ a filtering of graph pairs prior to the actual graph matching. To this end, we estimate the dissimilarity of graphs by means of the distribution of nodes or edges in a polar coordinate system termed *polar graph dissimilarity* (PGD) (we denote these graph dissimilarities by PGD-Node and PGD-Edge, respectively). Finally, BP is compared if, and only if, this estimated dissimilarity is below a certain threshold. We denote this filtering approach by BP-FRN and BP-FRE (depending on whether PGD-Node or PGD-Edge is used for filtering).

Rather than speeding up BP by means of filters and other heuristics, we also employ suboptimal algorithms for GED with quadratic — rather than cubic – time complexity such as for instance the *Hausdorff edit distance* (HED) [301]. HED reduces the complexity of GED to a set matching problem allowing us to have multiple node assignments. HED allows us to consider a larger structural node context — rather than adjacent edges only — due to its lower complexity [94]. As a result, the so-called *context-aware Hausdorff edit distance* (CED) can be employed in our framework. Finally, we make use of another suboptimal algorithm for GED with quadratic time complexity, namely *bipartite graph edit distance 2* (BP2). BP2 is based on a greedy-assignment of BP that makes use of the quadratic set assignment approach of HED [95].

In an exhaustive experimental evaluation on four different historical manuscripts, namely *George Washington* (GW), *Parzival* (PAR), *Alvermann Konzilsprotokolle* (AK), and *Botany* (BOT), we compare all the graph-based approaches with each other. First, we evaluate the different graph representations on the baseline system BP. We observe that either `Keypoint` or `Projection` graphs lead to the highest accuracy rates. Next, we empirically evaluate the different speed-up approaches with BP. By using BP-Q we observe speed-up factors between and 4 and 22 when compared with BP. However, this reduction in computation time is accompanied by a marginal decline in accuracy on most of the manuscripts. We observe similar speed-up factors for BP-FRN and BP-FRE. However, with these filters we also observe an improvement in overall KWS accuracy. Using the linear time graph dissimilarities PGD-Node and PGD-Edge as basic graph dissimilarities, we observe very high speed-up factors of up to 5,000 when compared with BP. However, in the case of PGD-Node, this speed-up is accompanied with a drop in KWS accuracy, while PGD-Edge is able to keep up with, or even outperform, the accuracy of BP.

In terms of the quadratic time algorithm HED, we observe both speed-up factors of up to 230 and an increase in accuracy when compared with BP. We assume that the multiple node assignments in fact allowed in HED allow a certain warping of character graphs, making HED more robust towards variations. KWS accuracy can be further improved by using CED rather than HED. However, one needs to consider that the speed-up factors of CED are declining with the increase of the context radius used in CED. By means of BP2, we observe similar or slightly better accuracy rates than with BP. Due to the quadratic time complexity of BP2 we observe similar speed-up factors as with HED.

Finally, we also developed and researched a number of different ensemble methods to combine graph edit distances stemming from different graph representations. We observe substantial improvements in overall KWS accuracy. In fact, ensemble methods lead to higher accuracy on all manuscripts than the corresponding individual methods. In particular, we observe that ensemble methods have a beneficial influence on manuscripts with large intraword variations as observed on BOT.

In order to proof the generalisability of the proposed KWS framework, a cross-evaluation on all four manuscripts is conducted. This cross-evaluation allows us to verify whether or not the proposed system can be employed for novel manuscripts without prior cost model optimisation. In particular, the optimised cost model of one manuscript (e.g. GW) is used to test all three remaining manuscripts (e.g. PAR, AK, and BOT). We observe that the employed cost model is directly dependent on the characteristics of the handwriting (i.e. density and flourish). Hence, the cost model for novel manuscripts can be directly derived based on the characteristics of the handwriting. We need not carry out an exhaustive parameter optimisation.

This book also provides a deep and broad comparison of our framework with different state-of-the-art template-based and learning-based reference systems. In the case of template-based approaches, we compare our framework with four *dynamic time warping* (DTW)-based reference systems on the GW and PAR manuscripts. We observe that HED and CED, but also BP-FRN and BP-FRE substantially outperform most template-based reference approaches. In particular, by using CED we observe the highest accuracy rates of all the approaches (not, however, considering the ensembles). This is quite astonishing as some of the evaluated DTW-based reference systems make use of advanced feature sets, while our graphs represent spatial information only. Moreover, one of the employed reference systems makes use of an unsupervised feature extraction approach by means of deep learning. Regarding the ensemble-based KWS methods proposed in this book, we observe further improvements when they are compared with the template-based reference system.

Graph-based approaches can keep up with DTW systems, not only in terms of accuracy, but also with respect to runtime. In particular, the quadratic time graph matching algorithms HED, CED, and BP2 achieve similar runtimes to the DTW-based approaches. When using the linear time algorithms PGD-Node and PGD-Edge, we observe the lowest matching times of all template-based approaches.

Last but not least, we compare the proposed graph-based KWS systems with three state-of-the-art learning-based approaches on the AK and BOT manuscripts. We observe that one *convolutional neural network* (CNN) approach clearly outperforms all the other approaches. In the case of single graph-based approaches, however, our framework can keep up with, or even outperform, the two other learning-based approaches on the AK manuscript. We observe the clear advantages of the evaluated learning-based approaches on the more challenging BOT manuscript. However, a different tendency in performance can be observed with our novel graph-based ensemble methods. That is, our HED-based and CED-based ensemble methods achieve very similar results to the learning-based systems. This is quite interesting, especially as we need to keep in mind that none of the graph-based approaches are dependent on labelled training data, while learning-based approaches are often crucially dependent on a typically large training set. In the case of handwritten historical documents, the acquisition of labelled training data is not only labour-intensive and costly, but also often inherently limited by the size of the document.

As a final observation, we conclude that graph-based approaches for KWS offer manifold possibilities, advantages, and research opportunities. In particular, this book shows how various graph representations can be extracted from word images and researches versatile graph matching approaches with cubic, quadratic, and linear time complexity. In an exhaustive experimental evaluation the different representations and matching schemes are thoroughly evaluated and compared for four benchmark datasets. We clearly show that our novel approaches are able to outperform state-of-the-art template-based approaches in terms of both accuracy and runtime. Moreover, this book shows that graph-based approaches to KWS are also capable of keeping up or even outperforming state-of-the-art learning-based KWS.

8.2 Future Work

A limitation of the proposed KWS framework might be the need for a full segmentation of document pages into isolated word images. This does not limit the applicability of our research but leads to an additional processing step (which might produce errors). Technically speaking, two approaches for a segmentation-free KWS framework in conjunction with graphs would be possible. First, a complete document or lines of documents could be represented by means of large graphs, and thus, keywords could then be retrieved, for instance, by means of subgraph isomorphism [208, 209, 215]. Second, candidate word images (so-called *word hypotheses*) could be detected automatically by means of various approaches [170]. This would allow a seamless integration of the proposed KWS framework towards a segmentation-free approach.

Moreover, the researched graph-based KWS framework is limited to template-based KWS. That is, the query is limited to visual inputs (known as *query-by-example* (QbE)), and thus, textual queries (termed *query-by-string* (QbS)) are not supported. In order to bridge this gap, one could, for instance, generate a graph synthetically on the basis of a textual query. Another possibility would be the use of graph kernels [200–202] and graph embedding [197–199] in the context of KWS. This would allow us to employ more sophisticated learning-based approaches that might provide seamless integration of query-by-string techniques.

In addition to these two limitations, we see great potential for other lines of research with respect to our graph-based KWS system. One idea would be to use graph embedding techniques in conjunction with DTW.

Roughly speaking, a subgraph could be extracted at every position of a sliding window on the word images. These subgraphs can then either be matched against other subgraphs of a document graph or a set of prototype graphs (representing common n-grams). The resulting dissimilarity matrix could then be used to find an optimal alignment between words by means of DTW. However, a limitation of this approach might be the high sensitivity of DTW towards scaling. Consequently, one first needs to research novel scale-invariant graph representations (for instance, Keypoint graphs with intermediate nodes that are based on the size of the word images).

Another rewarding research area might be the definition and research of novel graph representations for KWS. For example, one could research formalisms that make extensive use of edges, star-shaped graphs, or pyramidal graphs [302–306]. Moreover, one could also endow our framework with other graph-based and learning-based approaches. In particular, with the advent of deep learning approaches for graphs [179, 307–309], further improvements can be expected with respect to accuracy and querying. That is, these learning-based approaches would allow us to project graphs into a common subspace with textual representations (similar to the principle of word embedding). Hence, visual and textual queries could then be employed in our graph-based KWS framework.

Chapter 9

Visualisation of Graph Representations

The visualisation of the four different graph representation formalisms (i.e. Keypoint, Grid, Projection, and Split) is based on graphs with different mean numbers of nodes (i.e. small = 25-50 mean number of nodes, medium = 50-75 mean number of nodes, large = 75-100 mean number of nodes). That is, for every manuscript parameters have been chosen to meet the desired mean number of nodes, see Table 9.1 for *George Washington* (GW), Table 9.2 for *Parzival* (PAR), Table 9.3 for *Alvermann Konzilsprotokolle* (AK), and Table 9.4 for *Botany* (BOT).

In the following sections, five different word graphs in three different sizes are visualised for each manuscript and graph representation.

Table 9.1: GW: Parameters of graph representations for visualisation.

Method	Size	Parameter	
Keypoint	Small	$D = 16$	
	Medium	$D = 12$	
	Large	$D = 8$	
Grid	Small	$w = 14$	$h = 12$
	Medium	$w = 12$	$h = 10$
	Large	$w = 10$	$h = 8$
Projection	Small	$D_v = 8$	$D_h = 11$
	Medium	$D_v = 7$	$D_h = 9$
	Large	$D_v = 6$	$D_h = 7$
Split	Small	$D_w = 11$	$D_h = 12$
	Medium	$D_w = 9$	$D_h = 10$
	Large	$D_w = 7$	$D_h = 8$

Table 9.2: PAR: Parameters of graph representations for visualisation.

Method	Size	Parameter	
Keypoint	Small	$D = 8$	
	Medium	$D = 6$	
	Large	$D = 4$	
Grid	Small	$w = 8$	$h = 7$
	Medium	$w = 7$	$h = 6$
	Large	$w = 6$	$h = 5$
Projection	Small	$D_v = 10$	$D_h = 5$
	Medium	$D_v = 8$	$D_h = 4$
	Large	$D_v = 6$	$D_h = 3$
Split	Small	$D_w = 9$	$D_h = 5$
	Medium	$D_w = 7$	$D_h = 4$
	Large	$D_w = 5$	$D_h = 3$

Table 9.3: AK: Parameters of graph representations for visualisation.

Method	Size	Parameter	
Keypoint	Small	$D = 19$	
	Medium	$D = 16$	
	Large	$D = 13$	
Grid	Small	$w = 21$	$h = 16$
	Medium	$w = 18$	$h = 14$
	Large	$w = 15$	$h = 12$
Projection	Small	$D_v = 18$	$D_h = 12$
	Medium	$D_v = 16$	$D_h = 10$
	Large	$D_v = 14$	$D_h = 8$
Split	Small	$D_w = 23$	$D_h = 16$
	Medium	$D_w = 19$	$D_h = 14$
	Large	$D_w = 15$	$D_h = 12$

Table 9.4: BOT: Parameters of graph representations for visualisation.

Method	Size	Parameter	
Keypoint	Small	$D = 26$	
	Medium	$D = 20$	
	Large	$D = 14$	
Grid	Small	$w = 20$	$h = 15$
	Medium	$w = 18$	$h = 13$
	Large	$w = 16$	$h = 11$
Projection	Small	$D_v = 18$	$D_h = 12$
	Medium	$D_v = 16$	$D_h = 10$
	Large	$D_v = 14$	$D_h = 8$
Split	Small	$D_w = 23$	$D_h = 16$
	Medium	$D_w = 19$	$D_h = 14$
	Large	$D_w = 15$	$D_h = 12$

9.1 Visualisation of GW

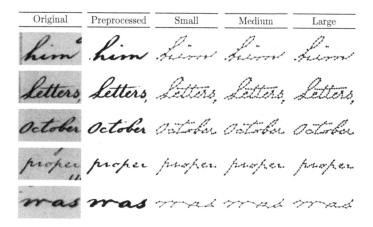

Original	Preprocessed	Small	Medium	Large

Fig. 9.1: GW: Original and preprocessed word images, as well as corresponding Keypoint graph representations.

Original
Preprocessed
Small
Medium
Large

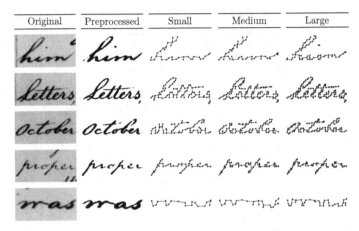

Fig. 9.2: GW: Original and preprocessed word images, as well as corresponding `Grid` graph representations.

Original
Preprocessed
Small
Medium
Large

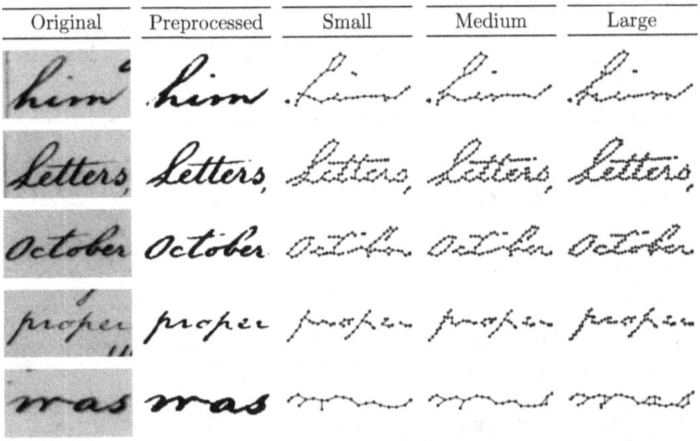

Fig. 9.3: GW: Original and preprocessed word images, as well as corresponding `Projection` graph representations.

Original	Preprocessed	Small	Medium	Large

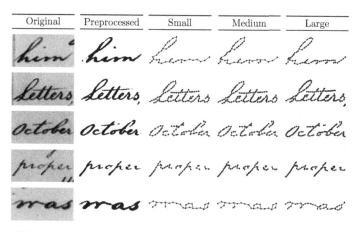

Fig. 9.4: GW: Original and preprocessed word images, as well as corresponding Split graph representations.

9.2 Visualisation of PAR

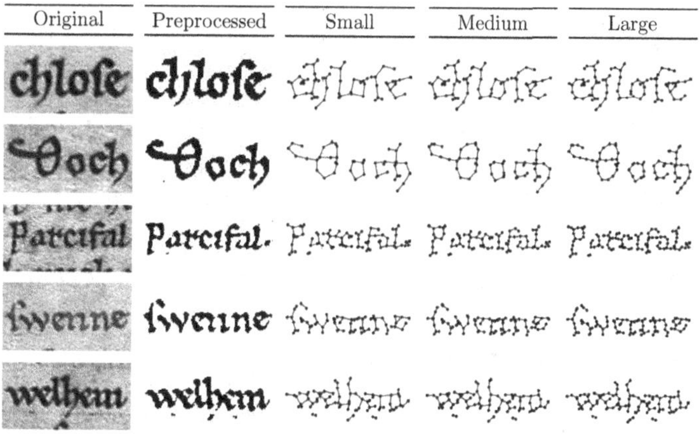

Original	Preprocessed	Small	Medium	Large

Fig. 9.5: PAR: Original and preprocessed word images, as well as corresponding Keypoint graph representations.

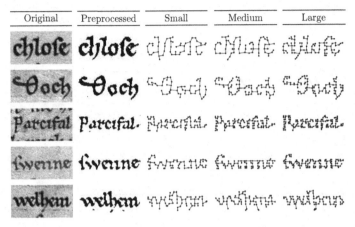

Original	Preprocessed	Small	Medium	Large

Fig. 9.6: PAR: Original and preprocessed word images, as well as corresponding `Grid` graph representations.

Original	Preprocessed	Small	Medium	Large

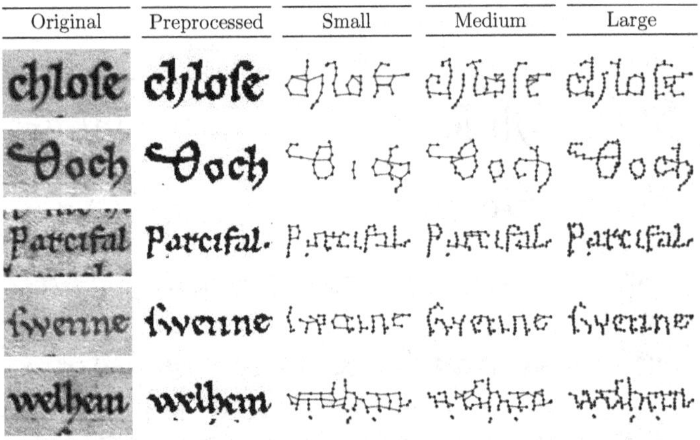

Fig. 9.7: PAR: Original and preprocessed word images, as well as corresponding `Projection` graph representations.

Original	Preprocessed	Small	Medium	Large

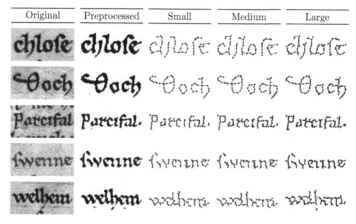

Fig. 9.8: PAR: Original and preprocessed word images, as well as corresponding Split graph representations.

9.3 Visualisation of AK

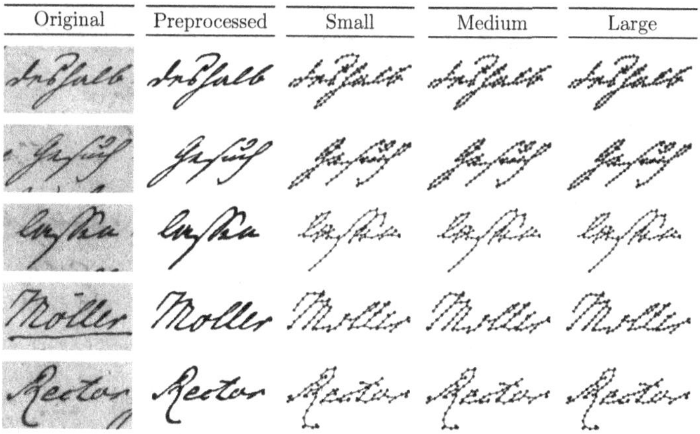

Original	Preprocessed	Small	Medium	Large

Fig. 9.9: AK: Original and preprocessed word images, as well as corresponding Keypoint graph representations.

Original
Preprocessed
Small
Medium
Large

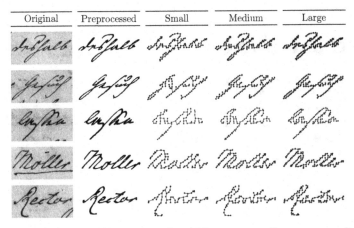

Fig. 9.10: AK: Original and preprocessed word images, as well as corresponding `Grid` graph representations.

Original
Preprocessed
Small
Medium
Large

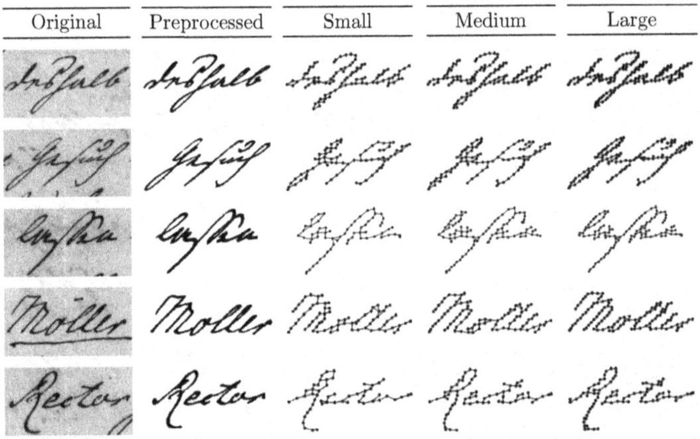

Fig. 9.11: AK: Original and preprocessed word images, as well as corresponding `Projection` graph representations.

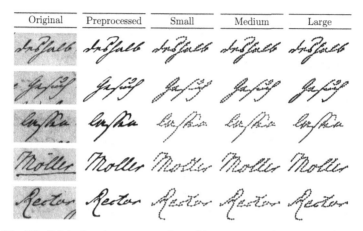

Fig. 9.12: AK: Original and preprocessed word images, as well as corresponding Split graph representations.

9.4 Visualisation of BOT

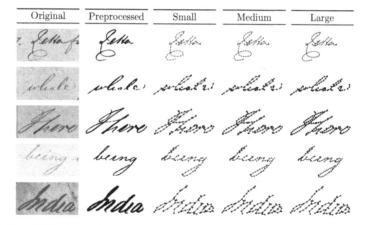

Fig. 9.13: BOT: Original and preprocessed word images, as well as corresponding Keypoint graph representations.

Original
Preprocessed
Small
Medium
Large

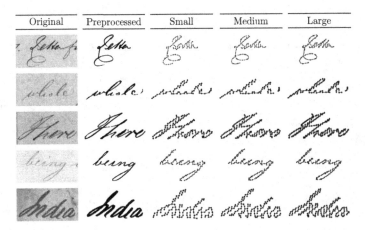

Fig. 9.14: BOT: Original and preprocessed word images, as well as corresponding `Grid` graph representations.

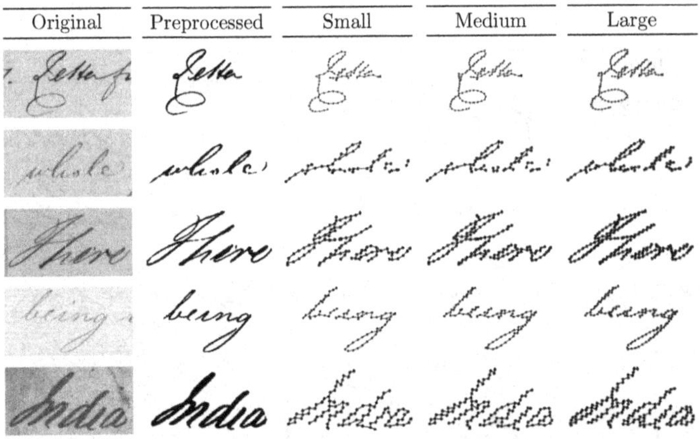
Original
Preprocessed
Small
Medium
Large

Fig. 9.15: BOT: Original and preprocessed word images, as well as corresponding `Projection` graph representations.

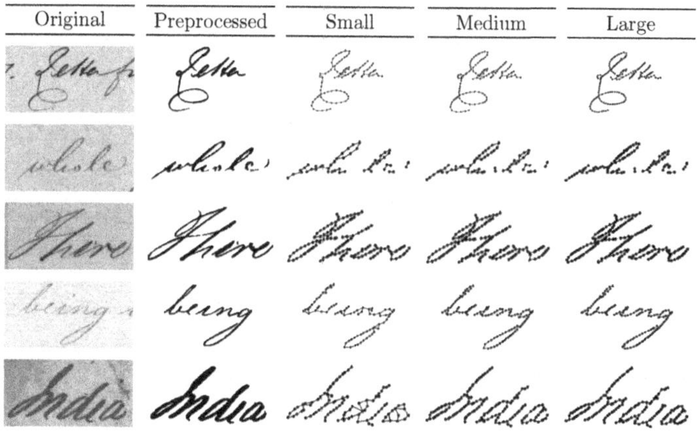

Original	Preprocessed	Small	Medium	Large

Fig. 9.16: BOT: Original and preprocessed word images, as well as corresponding Split graph representations.

Chapter 10

Optimisation of the Parameters

The optimisation of the parameters is conducted in six subsequent steps as described in Subsection 7.2.1. That is, for every manuscript (i.e. *George Washington* (GW), *Parzival* (PAR), *Alvermann Konzilsprotokolle* (AK), and *Botany* (BOT)) parameters are optimised individually. However, for *bipartite graph edit distance* (BP) only, all parameters are provided in Subsection 7.2.1, and thus, we provide the complete list of optimised parameters for all remaining graph matching algorithms (i.e. *Hausdorff edit distance* (HED), *context-aware Hausdorff edit distance* (CED), *bipartite graph edit distance 2* (BP2), and *polar graph dissimilarity* (PGD)) in the following sections. In Table 10.1, optimised parameters are provided for cost functions of *graph edit distance* (GED) computation, while the fast rejection threshold D is optimised with respect to the *filter rate* (FR) and *average precision* (AP) in Figs. 10.1, 10.2, 10.3, 10.4, 10.5 and 10.6. Finally, the retrieval index r_2 is optimised in Tables 10.2 and 10.3.

Table 10.1: Optimal cost function parameter for graph edit distance computation.

Method	GW				PAR				AK				BOT			
	τ_v	τ_e	α	β	τ_v	τ_e	α	β	τ_v	τ_e	α	β	τ_v	τ_e	α	β
HED																
Keypoint	8	4	0.5	0.1	8	1	0.1	0.5	16	1	0.3	0.1	8	4	0.3	0.1
Grid	4	1	0.9	0.1	4	8	0.7	0.5	-	-	-	-	-	-	-	-
Projection	4	1	0.1	0.1	8	1	0.1	0.7	1	4	0.1	0.1	8	32	0.5	0.1
Split	4	8	0.5	0.1	4	1	0.1	0.5	-	-	-	-	-	-	-	-
CED ($n = 2$)																
Keypoint	8	4	0.5	0.1	8	4	0.3	0.5	4	16	0.3	0.3	4	16	0.5	0.1
Grid	8	1	0.3	0.1	4	8	0.7	0.5	-	-	-	-	-	-	-	-
Projection	8	1	0.3	0.1	4	4	0.5	0.5	32	4	0.1	0.1	16	32	0.5	0.1
Split	8	1	0.1	0.1	8	1	0.1	0.5	-	-	-	-	-	-	-	-
CED ($n = 4$)																
Keypoint	8	4	0.5	0.1	4	1	0.1	0.5	32	16	0.7	0.3	8	4	0.1	0.1
Grid	8	1	0.7	0.1	4	4	0.5	0.5	-	-	-	-	-	-	-	-
Projection	8	1	0.3	0.1	4	4	0.5	0.3	32	16	0.5	0.1	32	4	0.1	0.1
Split	8	4	0.5	0.1	8	4	0.5	0.3	-	-	-	-	-	-	-	-
CED ($n = 6$)																
Keypoint	16	8	0.7	0.1	4	1	0.1	0.3	8	4	0.1	0.3	32	8	0.3	0.1
Grid	8	1	0.3	0.1	8	8	0.9	0.5	-	-	-	-	-	-	-	-
Projection	8	4	0.5	0.1	8	4	0.5	0.3	8	32	0.5	0.1	8	16	0.3	0.1
Split	8	1	0.1	0.1	4	8	0.7	0.3	-	-	-	-	-	-	-	-
BP2																
Keypoint	8	4	0.7	0.1	8	4	0.5	0.3	32	4	0.3	0.3	16	4	0.3	0.3
Grid	4	4	0.7	0.1	4	8	0.9	0.7	-	-	-	-	-	-	-	-
Projection	4	1	0.3	0.1	4	1	0.5	0.5	8	8	0.5	0.1	8	8	0.9	0.1
Split	4	1	0.7	0.1	4	1	0.9	0.5	-	-	-	-	-	-	-	-

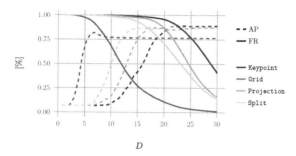

Fig. 10.1: PAR: *AP* and *FR* for BP-FRN as a function of threshold *D*.

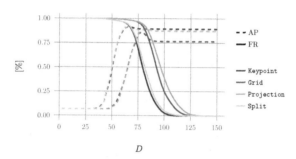

Fig. 10.2: PAR: *AP* and *FR* for BP-FRE as a function of threshold *D*.

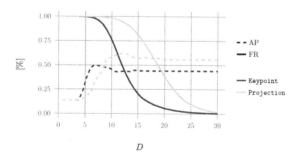

Fig. 10.3: AK: *AP* and *FR* for BP-FRN as a function of threshold *D*.

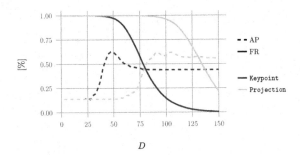

Fig. 10.4: AK: *AP* and *FR* for BP-FRE as a function of threshold *D*.

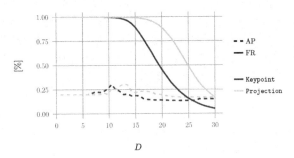

Fig. 10.5: BOT: *AP* and *FR* for BP-FRN as a function of threshold *D*.

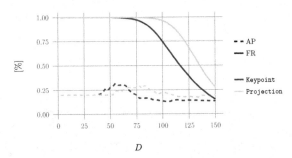

Fig. 10.6: BOT: *AP* and *FR* for BP-FRE as a function of threshold *D*.

Table 10.2: Optimal m for retrieval index r_2 using single graph-matching algorithms.

Method		GW	PAR	AK	BOT
BP-Q	Keypoint	4.10	2.30	0.80	2.05
	Projection	4.55	2.80	1.70	1.05
PGD-Node	Keypoint	0.330	0.025	0.190	0.060
	Grid	0.650	0.070	-	-
	Projection	0.205	0.020	0.095	0.040
	Split	0.230	0.045	-	-
PGD-Edge	Keypoint	0.030	0.005	0.015	0.015
	Grid	0.025	0.010	-	-
	Projection	0.035	0.005	0.010	0.015
	Split	0.020	0.005	-	-
HED	Keypoint	22.55	49.60	20.50	11.25
	Grid	8.00	14.25	-	-
	Projection	31.25	49.25	11.20	13.40
	Split	18.95	47.05	-	-
CED $(n = 2)$	Keypoint	20.60	20.70	9.70	10.15
	Grid	30.55	14.35	-	-
	Projection	28.40	13.90	48.50	11.60
	Split	49.95	47.65	-	-
CED $(n = 4)$	Keypoint	14.35	24.30	11.65	13.25
	Grid	12.45	8.10	-	-
	Projection	18.70	11.20	12.65	33.70
	Split	13.25	10.15	-	-
CED $(n = 6)$	Keypoint	18.10	20.75	9.35	13.30
	Grid	20.85	11.65	-	-
	Projection	12.80	15.15	8.40	11.35
	Split	35.50	12.65	-	-
BP2	Keypoint	8.45	3.80	6.40	4.50
	Grid	4.90	7.20	-	-
	Projection	6.00	5.60	2.65	4.50
	Split	4.30	4.70	-	-

Graph-Based Keyword Spotting

Table 10.3: Optimal m for retrieval index r_2 using ensemble methods.

Method		GW	PAR	AK	BOT
HED	min	32.10	45.95	19.45	14.90
	max	21.45	49.50	12.85	12.10
	mean	10.55	41.55	7.75	3.30
	sum_γ	34.40	45.60	16.80	16.60
	sum_{map}	15.40	47.95	10.25	11.65
CED (min)	$n = 2$	24.05	36.95	30.05	8.50
	$n = 4$	22.25	22.85	13.35	27.15
	$n = 6$	13.40	21.80	12.75	10.70
	$m = \text{K}$	18.45	40.35	13.00	13.40
	$m = \text{P}$	24.05	18.95	48.65	22.55
	All	24.05	36.95	48.65	22.55
CED (max)	$n = 2$	14.45	19.20	7.00	2.05
	$n = 4$	11.95	16.50	9.45	13.30
	$n = 6$	11.35	19.00	8.65	7.55
	$m = \text{K}$	14.85	21.75	10.15	2.05
	$m = \text{P}$	12.35	13.95	8.65	8.30
	All	11.70	13.95	9.25	4.25
CED (mean)	$n = 2$	6.80	35.85	10.15	1.85
	$n = 4$	6.25	30.05	5.95	8.55
	$n = 6$	5.80	32.20	4.90	3.80
	$m = \text{K}$	3.95	10.35	3.90	2.55
	$m = \text{P}$	5.25	13.95	4.35	4.85
	All	2.35	4.05	2.00	1.85
CED (sum_γ)	$n = 2$	18.20	19.95	47.20	5.75
	$n = 4$	16.25	14.60	15.40	21.40
	$n = 6$	11.45	17.70	8.85	9.35
CED (sum_{map})	$n = 2$	13.60	16.55	10.80	3.60
	$n = 4$	12.45	12.20	10.00	16.40
	$n = 6$	10.55	15.65	5.25	4.85
BP2	min	6.10	8.90	8.25	3.35
	max	3.85	6.05	3.40	2.55
	mean	3.40	6.25	2.45	1.60
	sum_γ	5.15	8.10	6.00	4.05
	sum_{map}	3.80	6.30	2.75	3.10

Chapter 11

Ensemble Methods

Ensemble methods for graph-based *keyword spotting* (KWS) allow us to combine the *graph edit distances* (GEDs) of different graph representations of the same manuscript (i.e. *George Washington* (GW), *Parzival* (PAR), *Alvermann Konzilsprotokolle* (AK), and *Botany* (BOT)) as introduced in Section 6.3.3. In Section 7.7.1, experiments for ensemble methods by means of *bipartite graph edit distance* (BP) are discussed, while results of ensemble methods for *Hausdorff edit distance* (HED), *context-aware Hausdorff edit distance* (CED), and *bipartite graph edit distance 2* (BP2) are summarised in Section 7.7.2. The remaining results (i.e. *mean average precision* (MAP) and *average precision* (AP) for HED, CED, and BP2) are provided in the following sections. In Section 11.1, results for HED-based ensemble methods are provided, while CED-based ensemble methods are presented in Section 11.2. Finally, ensemble methods based on BP2 are presented in Section 11.3.

11.1   Ensemble Methods Based on HED

Table 11.1: GW: MAP and AP of the KWS system using different HED-based ensemble methods. With ± we indicate the relative percental gain or loss in the accuracy of the ensemble methods when compared with HED. The best result per column is shown in bold.

Method	MAP	±	AP	±
		Single		
HED	69.28		57.38	
		Ensemble		
min	68.83	−0.44	57.36	−0.02
max	71.36	+2.08	60.44	+3.05
mean	**73.64**	+4.37	**63.84**	+6.46
sum$_\gamma$	71.82	+2.54	60.49	+3.10
sum$_{map}$	73.58	+4.31	**63.84**	+6.46

Table 11.2: PAR: MAP and AP of the KWS system using different HED-based ensemble methods. With ± we indicate the relative percental gain or loss in the accuracy of the ensemble methods when compared with HED. The best result per column is shown in bold.

Method	MAP	±	AP	±
		Single		
HED	72.82		76.69	
		Ensemble		
min	79.99	+7.17	84.07	+7.38
max	76.21	+3.39	83.88	+7.19
mean	80.46	+7.64	**88.37**	+11.68
sum$_\gamma$	79.03	+6.21	85.24	+8.55
sum$_{map}$	**80.58**	+7.76	87.98	+11.29

Table 11.3: AK: MAP and AP of the KWS system using different HED-based ensemble methods. With ± we indicate the relative percental gain or loss in the accuracy of the ensemble methods when compared with HED. The best result per column is shown in bold.

Method	MAP	±	AP	±
		Single		
HED	81.06		79.81	
		Ensemble		
min	80.40	−0.66	77.82	−1.99
max	83.65	+2.59	80.15	+0.34
mean	83.42	+2.36	80.90	+1.09
sum$_\gamma$	**83.80**	+2.74	80.21	+0.40
sum$_{map}$	83.44	+2.38	**80.97**	+1.16

Table 11.4: BOT: MAP and AP of the KWS system using different HED-based ensemble methods. With ± we indicate the relative percental gain or loss in the accuracy of the ensemble methods when compared with HED. The best result per column is shown in bold.

Method	MAP	±	AP	±
		Single		
HED	51.74		40.29	
		Ensemble		
min	67.99	+16.25	45.57	+5.28
max	**71.00**	+19.26	**49.88**	+9.59
mean	70.67	+18.93	49.13	+8.84
sum$_\gamma$	69.69	+17.95	48.15	+7.86
sum$_{map}$	70.67	+18.93	49.17	+8.88

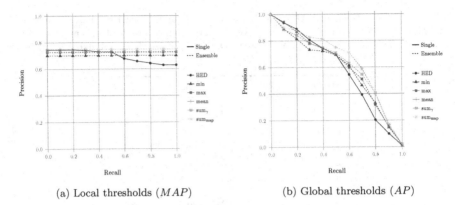

(a) Local thresholds (MAP)   (b) Global thresholds (AP)

Fig. 11.1: GW: Recall-precision curves using HED and HED-based ensemble methods.

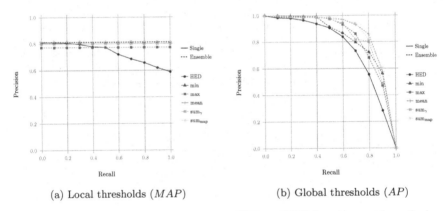

(a) Local thresholds (MAP)   (b) Global thresholds (AP)

Fig. 11.2: PAR: Recall-precision curves using HED and HED-based ensemble methods.

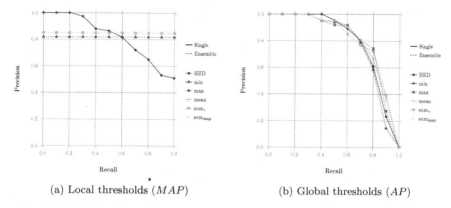

(a) Local thresholds (MAP) (b) Global thresholds (AP)

Fig. 11.3: AK: Recall-precision curves using HED and HED-based ensemble methods.

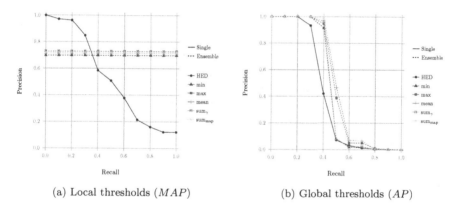

(a) Local thresholds (MAP) (b) Global thresholds (AP)

Fig. 11.4: BOT: Recall-precision curves using HED and HED-based ensemble methods.

11.2 Ensemble Methods Based on CED

Table 11.5: GW: MAP and AP of the KWS system using different CED-based ensemble methods. With ± we indicate the relative percental gain or loss in the accuracy of the ensemble methods when compared with CED. The best result per column is shown in bold.

Method	MAP	±	AP	±
Single				
CED	71.23		59.15	
Ensemble min				
Graph $n = 2$	70.00	−1.23	58.96	−0.18
Graph $n = 4$	67.02	−4.21	55.38	−3.76
Graph $n = 6$	68.70	−2.53	59.67	+0.52
Context $m = $ K	74.14	+2.91	64.15	+5.00
Context $m = $ P	69.60	−1.63	58.52	−0.63
All	70.47	−0.76	59.10	−0.05
Ensemble max				
Graph $n = 2$	74.01	+2.78	64.42	+5.27
Graph $n = 4$	72.50	+1.27	63.17	+4.02
Graph $n = 6$	63.06	−8.18	51.00	−8.15
Context $m = $ K	72.81	+1.58	61.71	+2.56
Context $m = $ P	63.05	−8.18	50.98	−8.17
All	63.72	−7.51	52.51	−6.64
Ensemble mean				
Graph $n = 2$	74.62	+3.39	64.74	+5.59
Graph $n = 4$	73.22	+1.99	63.49	+4.35
Graph $n = 6$	69.67	−1.56	58.23	−0.92
Context $m = $ K	74.27	+3.04	64.27	+5.12
Context $m = $ P	67.35	−3.88	56.54	−2.60
All	73.06	+1.83	63.73	+4.58
Ensemble sum$_\gamma$				
Graph $n = 2$	73.32	+2.09	64.10	+4.95
Graph $n = 4$	71.67	+0.44	62.03	+2.88
Graph $n = 6$	66.83	−4.40	55.65	−3.50
Ensemble sum$_{map}$				
Graph $n = 2$	**74.71**	+3.48	**65.21**	+6.07
Graph $n = 4$	73.23	+2.00	63.43	+4.29
Graph $n = 6$	69.56	−1.67	58.44	−0.71

Table 11.6: PAR: MAP and AP of the KWS system using different CED-based ensemble methods. With ± we indicate the relative percental gain or loss in the accuracy of the ensemble methods when compared with CED. The best result per column is shown in bold.

Method	MAP	±	AP	±
	Single			
CED	74.10		77.94	
	Ensemble min			
Graph $n = 2$	73.29	−0.81	80.48	+2.54
Graph $n = 4$	74.65	+0.55	80.45	+2.51
Graph $n = 6$	73.57	−0.53	80.15	+2.21
Context $m = $ K	74.79	+0.69	82.88	+4.94
Context $m = $ P	77.46	+3.36	81.69	+3.75
All	74.79	+0.69	83.02	+5.08
	Ensemble max			
Graph $n = 2$	77.77	+3.67	76.91	−1.03
Graph $n = 4$	78.62	+4.52	78.19	+0.25
Graph $n = 6$	77.39	+3.29	76.95	−0.99
Context $m = $ K	73.59	−0.51	80.63	+2.69
Context $m = $ P	78.62	+4.52	85.02	+7.08
All	78.62	+4.52	85.02	+7.08
	Ensemble mean			
Graph $n = 2$	80.06	+5.96	80.88	+2.94
Graph $n = 4$	79.79	+5.69	80.63	+2.69
Graph $n = 6$	79.11	+5.01	80.77	+2.83
Context $m = $ K	77.36	+3.26	84.10	+6.16
Context $m = $ P	78.66	+4.56	79.30	+1.36
All	**81.08**	+6.98	86.50	+8.56
	Ensemble sum$_\gamma$			
Graph $n = 2$	78.67	+4.57	85.77	+7.83
Graph $n = 4$	78.90	+4.80	84.70	+6.76
Graph $n = 6$	78.53	+4.43	82.82	+4.88
	Ensemble sum$_{map}$			
Graph $n = 2$	80.19	+6.09	**86.94**	+9.00
Graph $n = 4$	79.93	+5.83	86.71	+8.77
Graph $n = 6$	78.96	+4.86	84.54	+6.60

Table 11.7: AK: MAP and AP of the KWS system using different CED-based ensemble methods. With $\pm$ we indicate the relative percental gain or loss in the accuracy of the ensemble methods when compared with CED. The best result per column is shown in bold.

Method	MAP	$\pm$	AP	$\pm$
		Single		
CED	82.69		81.46	
		Ensemble min		
Graph $n = 2$	82.26	−0.43	76.28	−5.18
Graph $n = 4$	83.35	+0.66	81.33	−0.13
Graph $n = 6$	83.93	+1.24	81.90	+0.44
Context $m = $ K	81.12	−1.57	77.63	−3.83
Context $m = $ P	82.26	−0.43	79.30	−2.16
All	82.26	−0.43	79.30	−2.16
		Ensemble max		
Graph $n = 2$	80.58	−2.11	76.13	−5.33
Graph $n = 4$	81.07	−1.62	77.83	−3.63
Graph $n = 6$	83.86	+1.17	80.31	−1.15
Context $m = $ K	82.73	+0.04	79.97	−1.49
Context $m = $ P	84.59	+1.90	82.39	+0.93
All	83.80	+1.11	80.04	−1.42
		Ensemble mean		
Graph $n = 2$	81.52	−1.17	79.05	−2.41
Graph $n = 4$	84.22	+1.53	80.02	−1.44
Graph $n = 6$	85.20	+2.51	82.62	+1.16
Context $m = $ K	83.42	+0.73	79.52	−1.94
Context $m = $ P	**85.46**	+2.77	82.75	+1.29
All	84.49	+1.80	82.04	+0.58
		Ensemble sum$_\gamma$		
Graph $n = 2$	84.18	+1.49	81.59	+0.13
Graph $n = 4$	83.37	+0.68	82.00	+0.54
Graph $n = 6$	85.42	+2.73	**83.29**	+1.83
		Ensemble sum$_{\text{map}}$		
Graph $n = 2$	81.52	−1.17	78.94	−2.52
Graph $n = 4$	84.22	+1.53	79.47	−1.99
Graph $n = 6$	85.20	+2.51	82.87	+1.41

Table 11.8: BOT: MAP and AP of the KWS system using different CED-based ensemble methods. With ± we indicate the relative percental gain or loss in the accuracy of the ensemble methods when compared with CED. The best result per column is shown in bold.

Method	MAP	±	AP	±
	Single			
CED	52.45		40.34	
	Ensemble min			
Graph $n = 2$	67.64	+15.19	45.51	+5.17
Graph $n = 4$	67.32	+14.87	45.06	+4.72
Graph $n = 6$	71.43	+18.98	48.08	+7.74
Context $m =$ K	71.54	+19.09	48.51	+8.17
Context $m =$ P	67.32	+14.87	44.86	+4.52
All	67.32	+14.87	44.86	+4.52
	Ensemble max			
Graph $n = 2$	70.82	+18.37	**50.33**	+9.99
Graph $n = 4$	71.42	+18.97	50.29	+9.95
Graph $n = 6$	65.61	+13.16	44.16	+3.82
Context $m =$ K	70.82	+18.37	50.22	+9.88
Context $m =$ P	65.61	+13.16	44.19	+3.85
All	67.34	+14.89	45.68	+5.34
	Ensemble mean			
Graph $n = 2$	71.83	+19.38	49.84	+9.50
Graph $n = 4$	72.43	+19.98	50.03	+9.69
Graph $n = 6$	69.17	+16.72	45.50	+5.16
Context $m =$ K	**72.51**	+20.06	50.19	+9.85
Context $m =$ P	67.60	+15.15	45.11	+4.77
All	71.67	+19.22	49.02	+8.68
	Ensemble sum_γ			
Graph $n = 2$	70.67	+18.22	50.02	+9.68
Graph $n = 4$	69.16	+16.71	46.20	+5.86
Graph $n = 6$	65.59	+13.14	44.58	+4.24
	Ensemble $\mathrm{sum}_{\mathrm{map}}$			
Graph $n = 2$	71.83	+19.38	49.85	+9.51
Graph $n = 4$	72.43	+19.98	50.03	+9.69
Graph $n = 6$	69.25	+16.80	44.85	+4.51

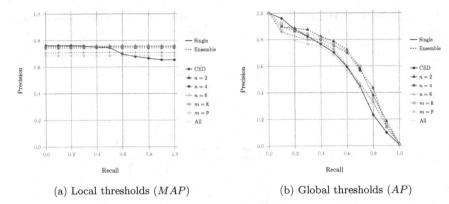

(a) Local thresholds (MAP) (b) Global thresholds (AP)

Fig. 11.5: GW: Recall-precision curves using CED and CED-based ensemble method mean.

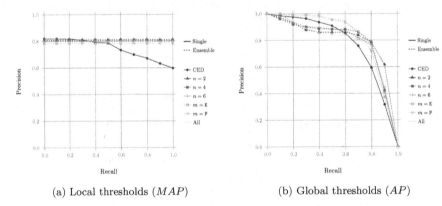

(a) Local thresholds (MAP) (b) Global thresholds (AP)

Fig. 11.6: PAR: Recall-precision curves using CED and CED-based ensemble method mean.

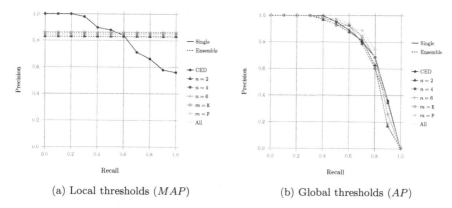

(a) Local thresholds (MAP) (b) Global thresholds (AP)

Fig. 11.7: AK: Recall-precision curves using CED and CED-based ensemble method mean.

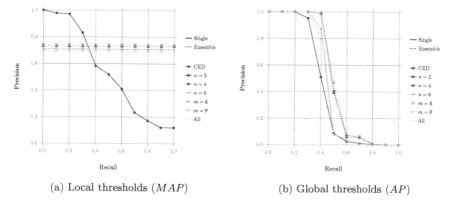

(a) Local thresholds (MAP) (b) Global thresholds (AP)

Fig. 11.8: BOT: Recall-precision curves using CED and CED-based ensemble method mean.

11.3   Ensemble Methods Based on BP2

Table 11.9: GW: MAP and AP of the KWS system using different BP2-based ensemble methods. With ± we indicate the relative percental gain or loss in the accuracy of the ensemble methods when compared with BP2. The best result per column is shown in bold.

Method	MAP	±	AP	±
		Single		
BP2	68.42		52.47	
		Ensemble		
min	**72.39**	+3.98	60.88	+8.40
max	68.47	+0.05	55.93	+3.46
mean	72.18	+3.77	62.38	+9.90
sum_γ	69.76	+1.34	58.53	+6.06
sum_map	71.93	+3.51	**63.32**	+10.85

Table 11.10: PAR: MAP and AP of the KWS system using different BP2-based ensemble methods. With ± we indicate the relative percental gain or loss in the accuracy of the ensemble methods when compared with BP2. The best result per column is shown in bold.

Method	MAP	±	AP	±
		Single		
BP2	68.69		69.54	
		Ensemble		
min	65.53	−3.16	69.57	+0.03
max	68.87	+0.18	78.84	+9.30
mean	74.77	+6.08	77.02	+7.48
sum_γ	74.73	+6.04	81.03	+11.49
sum_map	**74.82**	+6.13	**81.37**	+11.83

Table 11.11: AK: *MAP* and *AP* of the KWS system using different BP2-based ensemble methods. With ± we indicate the relative percental gain or loss in the accuracy of the ensemble methods when compared with BP2. The best result per column is shown in bold.

Method	MAP	±	AP	±
Single				
BP2	75.46		73.85	
Ensemble				
min	75.66	+0.20	72.35	−1.50
max	75.31	−0.15	76.23	+2.38
mean	**78.73**	+3.27	78.83	+4.98
sum_γ	77.83	+2.37	77.29	+3.44
sum_{map}	**78.73**	+3.27	**78.84**	+4.99

Table 11.12: BOT: *MAP* and *AP* of the KWS system using different BP2-based ensemble methods. With ± we indicate the relative percental gain or loss in the accuracy of the ensemble methods when compared with BP2. The best result per column is shown in bold.

Method	MAP	±	AP	±
Single				
BP2	50.94		39.25	
Ensemble				
min	**70.99**	+20.05	50.78	+11.53
max	67.49	+16.55	47.36	+8.11
mean	69.32	+18.38	49.98	+10.73
sum_γ	70.93	+19.99	**50.79**	+11.54
sum_{map}	69.32	+18.38	49.99	+10.74

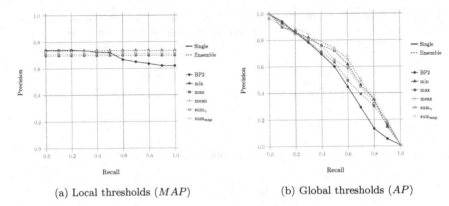

(a) Local thresholds (MAP)   (b) Global thresholds (AP)

Fig. 11.9: GW: Recall-precision curves using BP2 and BP2-based ensemble methods.

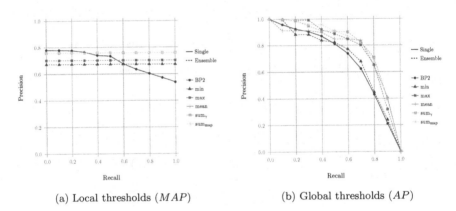

(a) Local thresholds (MAP)   (b) Global thresholds (AP)

Fig. 11.10: PAR: Recall-precision curves using BP2 and BP2-based ensemble methods.

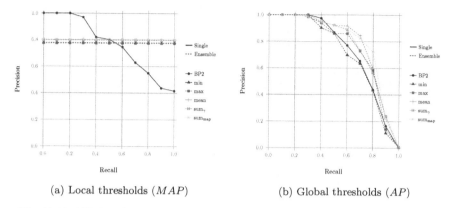

(a) Local thresholds (MAP)　　　　　(b) Global thresholds (AP)

Fig. 11.11: AK: Recall-precision curves using BP2 and BP2-based ensemble methods.

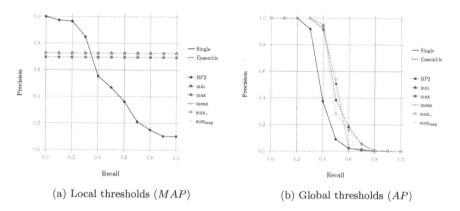

(a) Local thresholds (MAP)　　　　　(b) Global thresholds (AP)

Fig. 11.12: BOT: Recall-precision curves using BP2 and BP2-based ensemble methods.

Bibliography

[1] R. O. Duda, P. E. Hart and D. G. Stork, *Pattern Classification*. Wiley (2000).

[2] C. M. Bishop, Pattern Recognition and Machine Learning, *Pattern Recognition* **4**, 4, p. 738 (2006).

[3] O. Russakovsky, J. Deng, H. Su, J. Krause, S. Satheesh, S. Ma, Z. Huang, A. Karpathy, A. Khosla, M. Bernstein, A. C. Berg and L. Fei-Fei, ImageNet Large Scale Visual Recognition Challenge, *International Journal of Computer Vision* **115**, 3, pp. 211–252 (2015).

[4] K. He, X. Zhang, S. Ren and J. Sun, Delving Deep into Rectifiers: Surpassing Human-level Performance on ImageNet Classification, in *International Conference on Computer Vision*. IEEE, pp. 1026–1034 (2015).

[5] A. Esteva, B. Kuprel, R. A. Novoa, J. Ko, S. M. Swetter, H. M. Blau and S. Thrun, Dermatologist-level Classification of Skin Cancer with Deep Neural Networks, *Nature* **542**, 7639, pp. 115–118 (2017).

[6] H. Bunke and P. S. P. Wang, *Handbook of Character Recognition and Document Image Analysis*. World Scientific (1997).

[7] R. Plamondon and S. Srihari, Online and Off-line Handwriting Recognition: A Comprehensive Survey, *Transactions on Pattern Analysis and Machine Intelligence* **22**, 1, pp. 63–84 (2000).

[8] H. Bunke, Recognition of Cursive Roman Handwriting - Past, Present and Future, in *International Conference on Document Analysis and Recognition*. IEEE, pp. 448–459 (2003).

[9] S. Impedovo, More than Twenty Years of Advancements on Frontiers in Handwriting Recognition, *Pattern Recognition* **47**, 3, pp. 916–928 (2014).

[10] Q. Ye and D. Doermann, Text Detection and Recognition in Imagery: A Survey, *Transactions on Pattern Analysis and Machine Intelligence* **37**, 7, pp. 1480–1500 (2015).

[11] J. Edwards, Y. W. Teh, R. Bock, M. Maire, G. Vesom and D. A. Forsyth, Making Latin Manuscripts Searchable Using GHMMs, in *International Conference on Neural Information Processing Systems*, Vol. 17. MIT Press, pp. 385–392 (2004).

[12] A. Kaltenmeier, T. Caesar, J. Gloger and E. Mandler, Sophisticated Topology of Hidden Markov Models for Cursive Script Recognition, in *International Conference on Document Analysis and Recognition*. IEEE, pp. 139–142 (1993).

[13] S. N. Srihari, Recognition of Handwritten and Machine-printed Text for Postal Address Interpretation, *Pattern Recognition Letters* **14**, 4, pp. 291–302 (1993).

[14] A. Brakensiek and G. Rigoll, Handwritten Address Recognition Using Hidden Markov Models, in *Reading and Learning*. Springer, pp. 103–122 (2004).

[15] S. Impedovo, P. S. P. Wang and H. Bunke, *Automatic Bankcheck Processing, Series in Machine Perception and Artificial Intelligence*, Vol. 28. World Scientific (1997).

[16] R. Palacios, A. Gupta and P. S. Wang., Handwritten Bank check Recognition of Courtesy Amounts, *International Journal of Image and Graphics* **04**, 02, pp. 203–222 (2004).

[17] A.-M. Awal, H. Mouchère and C. Viard-Gaudin, Towards Handwritten Mathematical Expression Recognition, in *International Conference on Document Analysis and Recognition*. IEEE, pp. 1046–1050 (2009).

[18] F. Álvaro, J. A. Sánchez and J. M. Benedí, Unbiased Evaluation of Handwritten Mathematical Expression Recognition, in *International Conference on Frontiers in Handwriting Recognition*. IEEE, pp. 181–186 (2012).

[19] J. Zhang, J. Du, S. Zhang, D. Liu, Y. Hu, J. Hu, S. Wei and L. Dai, Watch, Attend and Parse: An End-to-end Neural Network-based Approach to Handwritten Mathematical Expression Recognition, *Pattern Recognition* **71**, pp. 196–206 (2017).

[20] M. Liwicki and H. Bunke, IAM-OnDB - An On-line English Sentence Database Acquired from Handwritten Text on a Whiteboard, in *International Conference on Document Analysis and Recognition*. IEEE, pp. 956–961 (2005).

[21] M. Liwicki, A. Graves, H. Bunke and J. Schmidhuber, A Novel Approach to On-Line Handwriting Recognition Based on Bidirectional Long Short-Term Memory Networks, in *International Conference on Document Analysis and Recognition*. IEEE, pp. 367–371 (2007).

[22] J. Geiger, J. Schenk, F. Wallhoff and G. Rigoll, Optimizing the Number of States for HMM-Based On-line Handwritten Whiteboard Recognition, in *International Conference on Frontiers in Handwriting Recognition*. IEEE, pp. 107–112 (2010).

[23] A. Fornés and G. Sánchez, Analysis and Recognition of Music Scores, in *Handbook of Document Image Processing and Recognition*. Springer, pp. 749–774 (2014).

[24] P. Riba, A. Fornés and J. Lladós, Towards the Alignment of Handwritten Music Scores, in *International Workshop on Graphics Recognition*. Springer, pp. 103–116 (2017).

[25] A. Baró, P. Riba and A. Fornés, Towards the Recognition of Compound Music Notes in Handwritten Music Scores, in *International Conference on Frontiers in Handwriting Recognition*. IEEE, pp. 465–470 (2016).

[26] A. Antonacopoulos and A. C. Downton, Special Issue on the Analysis of Historical Documents, *International Journal on Document Analysis and Recognition* **9**, 2-4, pp. 75–77 (2007).

[27] A. Fischer, *Handwriting Recognition in Historical Documents*, Ph.D. thesis, University of Bern (2012).

[28] J. I. Toledo, S. Dey, A. Fornés and J. Lladós, Handwriting Recognition by Attribute Embedding and Recurrent Neural Networks, in *International Conference on Document Analysis and Recognition*. IEEE, pp. 1038–1043 (2017).

[29] A. Garz, *A Human-Centered Approach to Structural Image Analysis for Complex Historical Manuscripts*, Ph.D. thesis, University of Fribourg (2017).

[30] D. Fernández-Mota, J. Almazán, N. Cirera, A. Fornés and J. Lladós, BH2M: The Barcelona Historical, Handwritten Marriages Database, in *International Conference on Pattern Recognition*. IEEE, pp. 256–261 (2014).

[31] A. Fischer, V. Frinken, A. Fornés and H. Bunke, Transcription Alignment of Latin Manuscripts Using Hidden Markov Models, in *Workshop on Historical Document Imaging and Processing*. ACM, p. 29 (2011).

[32] A. Fischer, A. Keller, V. Frinken and H. Bunke, Lexicon-free Handwritten Word Spotting Using Character HMMs, *Pattern Recognition Letters* **33**, 7, pp. 934–942 (2012).

[33] F. Le Bourgeois and H. Emptoz, DEBORA: Digital AccEss to BOoks of the RenAissance, *International Journal on Document Analysis and Recognition* **9**, 2-4, pp. 193–221 (2007).

[34] H. Balk and A. Conteh, IMPACT: Centre of Competence in Text Digitisation, in *Workshop on Historical Document Imaging and Processing*. ACM, pp. 155–160 (2011).

[35] J. A. Sánchez, G. Mühlberger, B. Gatos, P. Schofield, K. Depuydt, R. M. Davis, E. Vidal and J. de Does, TranScriptorium: A European Project on Handwritten Text Recognition, in *Symposium on Document Engineering*. ACM, pp. 227–228 (2013).

[36] V. Romero, V. Bosch, C. Hernández, E. Vidal and J. A. Sánchez, A Historical Document Handwriting Transcription End-to-end System, in *Iberian Conference on Pattern Recognition and Image Analysis*. Springer, pp. 149–157 (2017).

[37] P. Kahle, S. Colutto, G. Hackl and G. Muhlberger, Transkribus - A Service Platform for Transcription, Recognition and Retrieval of Historical Documents, in *International Conference on Document Analysis and Recognition*. IEEE, pp. 19–24 (2017).

[38] M. Baechler, A. Fischer, N. Naji, I. Rolf, H. Bunke and J. Savoy, HisDoc: Historical Document Analysis, Recognition, and Retrieval, in *Digital Humanities*. University of Hamburg (2012).

[39] A. Garz, N. Eichenberger, M. Liwicki and R. Ingold, HisDoc 2.0: Toward Computer-assisted Paleography, in *Conference on Natural Sciences and Technology in Manuscript Analysis*. University of Hamburg, pp. 7:19–28 (2014).

[40] A. Fischer, A. Keller, V. Frinken and H. Bunke, HMM-Based Word Spotting in Handwritten Documents Using Subword Models, in *International Conference on Pattern Recognition.* IEEE, pp. 3416–3419 (2010).

[41] V. Frinken, A. Fischer, M. Baumgartner and H. Bunke, Keyword Spotting for Self-training of BLSTM NN Based Handwriting Recognition Systems, *Pattern Recognition* **47**, 3, pp. 1073–1082 (2014).

[42] B. Wicht, A. Fischer and J. Hennebert, Keyword Spotting with Convolutional Deep Belief Networks and Dynamic Time Warping, in *International Conference on Artificial Neural Networks.* Springer, pp. 113–120 (2016).

[43] R. Rose and D. Paul, A Hidden Markov Model-Based Keyword Recognition System, in *International Conference on Acoustics, Speech and Signal Processing.* IEEE, pp. 129–132 (1990).

[44] S.-S. Kuo and O. E. Agazzi, Keyword Spotting in Poorly Printed Documents Using Pseudo 2-D Hidden Markov Models, *Transactions on Pattern Analysis and Machine Intelligence* **16**, 8, pp. 842–848 (1994).

[45] R. Manmatha, Chengfeng Han and E. Riseman, Word Spotting: A New Approach to Indexing Handwriting, in *Computer Society Conference on Computer Vision and Pattern Recognition.* IEEE, pp. 631–637 (1996).

[46] T. M. Rath and R. Manmatha, Word Image Matching Using Dynamic Time Warping, in *Computer Society Conference on Computer Vision and Pattern Recognition*, Vol. 2. IEEE, pp. II–521–II–527 (2003).

[47] G. L. Scott and H. C. Longuet-Higgins, An Algorithm for Associating the Features of Two Images, *Proceedings of the Royal Society B: Biological Sciences* **244**, 1309, pp. 21–26 (1991).

[48] B. Zhang, S. N. Srihari and C. Huang, Word Image Retrieval Using Binary Features, in *Document Recognition and Retrieval.* International Society for Optics and Photonics, pp. 45–54 (2003).

[49] T. M. Rath and R. Manmatha, Word Spotting for Historical Documents, *International Journal on Document Analysis and Recognition* **9**, 2-4, pp. 139–152 (2007).

[50] T. Adamek, N. E. O'Connor and A. F. Smeaton, Word Matching Using Single Closed Contours for Indexing Handwritten Historical Documents, *International Journal on Document Analysis and Recognition* **9**, 2-4, pp. 153–165 (2006).

[51] U.-V. Marti and H. Bunke, Using a Statistical Language Model to Improve the Performance of an HMM-based Cursive Handwriting Recognition Systems, *International Journal of Pattern Recognition and Artificial Intelligence* **15**, 01, pp. 65–90 (2001).

[52] J. A. Rodríguez-Serrano and F. Perronnin, Local Gradient Histogram Features for Word Spotting in Unconstrained Handwritten Documents, in *International Conference on Frontiers in Handwriting Recognition*, pp. 7–12 (2008).

[53] K. Terasawa and Y. Tanaka, Slit Style HOG Feature for Document Image Word Spotting, in *International Conference on Document Analysis and Recognition.* IEEE, pp. 116–120 (2009).

[54] A. Silberpfennig, L. Wolf, N. Dershowitz, S. Bhagesh and B. B. Chaudhuri,

Improving OCR for an Under-resourced Script Using Unsupervised Word-Spotting, in *International Conference on Document Analysis and Recognition*. IEEE, pp. 706–710 (2015).

[55] S. Dey, A. Nicolaou, J. Lladós and U. Pal, Local Binary Pattern for Word Spotting in Handwritten Historical Document, in *International Workshop on Structural, Syntactic, and Statistical Pattern Recognition*. Springer, pp. 574–583 (2016).

[56] B. Wicht, A. Fischer and J. Hennebert, Deep Learning Features for Handwritten Keyword Spotting, in *International Conference on Pattern Recognition*. IEEE, pp. 3434–3439 (2016).

[57] V. Frinken, A. Fischer, R. Manmatha and H. Bunke, A Novel Word Spotting Method Based on Recurrent Neural Networks, *Transactions on Pattern Analysis and Machine Intelligence* **34**, 2, pp. 211–224 (2012).

[58] V. Lavrenko, T. Rath and R. Manmatha, Holistic Word Recognition for Handwritten Historical Documents, in *International Workshop on Document Image Analysis for Libraries*. IEEE, pp. 278–287 (2004).

[59] J. Chan, C. Ziftci and D. Forsyth, Searching Off-line Arabic Documents, in *Computer Society Conference on Computer Vision and Pattern Recognition*, Vol. 2. IEEE, pp. 1455–1462 (2006).

[60] F. Perronnin and J. A. Rodríguez-Serrano, Fisher Kernels for Handwritten Word-spotting, in *International Conference on Document Analysis and Recognition*. IEEE, pp. 106–110 (2009).

[61] J. A. Rodríguez-Serrano and F. Perronnin, Handwritten Word-spotting Using Hidden Markov Models and Universal Vocabularies, *Pattern Recognition* **42**, 9, pp. 2106–2116 (2009).

[62] L. Rothacker, M. Rusiñol and G. a. Fink, Bag-of-features HMMs for Segmentation-free Word Spotting in Handwritten Documents, in *International Conference on Document Analysis and Recognition*. IEEE, pp. 1305–1309 (2013).

[63] S. Thomas, C. Chatelain, L. Heutte, T. Paquet and Y. Kessentini, A Deep HMM Model for Multiple Keywords Spotting in Handwritten Documents, *Pattern Analysis and Applications* **18**, 4, pp. 1003–1015 (2014).

[64] A. Sharma and Pramod Sankar K., Adapting Off-the-Shelf CNNs for Word Spotting & Recognition, in *International Conference on Document Analysis and Recognition*. IEEE, pp. 986–990 (2015).

[65] S. Sudholt and G. A. Fink, PHOCNet: A Deep Convolutional Neural Network for Word Spotting in Handwritten Documents, in *International Conference on Frontiers in Handwriting Recognition*. IEEE, pp. 277–282 (2016).

[66] T. Wilkinson and A. Brun, Semantic and Verbatim Word Spotting using Deep Neural Networks, in *International Conference on Frontiers in Handwriting Recognition*. IEEE, pp. 307–312 (2016).

[67] S. Sudholt and G. A. Fink, Evaluating Word String Embeddings and Loss Functions for CNN-based Word Spotting, in *International Conference on Document Analysis and Recognition*. IEEE, pp. 493–498 (2017).

[68] G. Lluis, M. Rusiñol and D. Karatzas, LSDE : Levenshtein Space Deep

Embedding for Query-by-string Word Spotting, in *International Conference on Document Analysis and Recognition*. IEEE, pp. 499–504 (2017).

[69] J. Almazán, A. Gordo, A. Fornés and E. Valveny, Word Spotting and Recognition with Embedded Attributes, *Transactions on Pattern Analysis and Machine Intelligence* **36**, 12, pp. 2552–2566 (2014a).

[70] J. Almazán, A. Gordo, A. Fornés and E. Valveny, Segmentation-free Word Spotting with Exemplar SVMs, *Pattern Recognition* **47**, 12, pp. 3967–3978 (2014b).

[71] M. Khayyat, L. Lam and C. Y. Suen, Learning-Based Word Spotting System for Arabic Handwritten Documents, *Pattern Recognition* **47**, 3, pp. 1021–1030 (2014).

[72] M. Rusiñol, D. Aldavert, R. Toledo and J. Lladós, Browsing Heterogeneous Document Collections by a Segmentation-free Word Spotting Method, in *International Conference on Document Analysis and Recognition*. IEEE, pp. 63–67 (2011).

[73] D. Aldavert, M. Rusiñol, R. Toledo and J. Lladós, Integrating Visual and Textual Cues for Query-by-string Word Spotting, in *International Conference on Document Analysis and Recognition*. IEEE, pp. 511–515 (2013).

[74] M. Rusiñol, D. Aldavert, R. Toledo and J. Lladós, Efficient Segmentation-free keyword Spotting in Historical Document Collections, *Pattern Recognition* **48**, 2, pp. 545–555 (2015).

[75] P. Wang, V. Eglin, C. Garcia, C. Largeron, J. Lladós and A. Fornés, A Novel Learning-Free Word Spotting Approach Based on Graph Representation, in *International Workshop on Document Analysis Systems*. IEEE, pp. 207–211 (2014).

[76] Q. A. Bui, M. Visani and R. Mullot, Unsupervised Word Spotting Using a Graph Representation Based on Invariants, in *International Conference on Document Analysis and Recognition*. IEEE, pp. 616–620 (2015).

[77] P. Riba, J. Lladós and A. Fornés, Handwritten Word Spotting by Inexact Matching of Grapheme Graphs, in *International Conference on Document Analysis and Recognition*. IEEE, pp. 781–785 (2015).

[78] K. Riesen, M. Neuhaus and H. Bunke, Bipartite Graph Matching for Computing the Edit Distance of Graphs, in *International Workshop on Graph-Based Representations in Pattern Recognition*, Vol. 6. Springer, pp. 1–12 (2007).

[79] A. Fischer, C. Y. Suen, V. Frinken, K. Riesen and H. Bunke, Approximation of Graph Edit Distance Based on Hausdorff Matching, *Pattern Recognition* **48**, 2, pp. 331–343 (2015).

[80] I. Pratikakis, K. Zagoris, B. Gatos, J. Puigcerver, A. H. Toselli and E. Vidal, ICFHR2016 Handwritten Keyword Spotting Competition (H-KWS 2016), in *International Conference on Frontiers in Handwriting Recognition*. IEEE, pp. 613–618 (2016).

[81] M. Stauffer, A. Fischer and K. Riesen, Searching and Browsing in Historical Documents—State of the Art and Novel Approaches for Template-Based Keyword Spotting, in *Business Information Systems and Technology 4.0*. Springer, pp. 97–211 (2018a).

[82] M. Stauffer, A. Fischer and K. Riesen, Keyword Spotting in Historical Handwritten Documents Based on Graph Matching, *Pattern Recognition* **81**, pp. 240–253 (2018b).

[83] M. Stauffer, A. Fischer and K. Riesen, Filters for Graph-Based Keyword Spotting in Historical Handwritten Documents, *Pattern Recognition Letters*, In Press (2018c).

[84] M. R. Ameri, M. Stauffer, K. Riesen, T. D. Bui and A. Fischer, Graph-Based Keyword Spotting in Historical Manuscripts Using Hausdorff Edit Distance, *Pattern Recognition Letters*, In Press (2018).

[85] M. Stauffer, A. Fischer and K. Riesen, Graph-Based Keyword Spotting in Historical Handwritten Documents, in *International Workshop on Structural, Syntactic, and Statistical Pattern Recognition*. Springer, pp. 564–573 (2016a).

[86] M. Stauffer, A. Fischer and K. Riesen, A Novel Graph Database for Handwritten Word Images, in *International Workshop on Structural, Syntactic, and Statistical Pattern Recognition*. Springer, pp. 553–563 (2016b).

[87] M. Stauffer, T. Tschachtli, A. Fischer and K. Riesen, A Survey on Applications of Bipartite Graph Edit Distance, in *International Workshop on Graph-Based Representations in Pattern Recognition*. Springer, pp. 242–252 (2017a).

[88] M. Stauffer, A. Fischer and K. Riesen, Speeding-Up Graph-Based Keyword Spotting in Historical Handwritten Documents, in *International Workshop on Graph-Based Representations in Pattern Recognition*. Springer, pp. 83–93 (2017b).

[89] M. R. Ameri, M. Stauffer, K. Riesen, T. D. Bui and A. Fischer, Keyword Spotting in Historical Documents Based on Handwriting Graphs and Hausdorff Edit Distance, in *International Graphonomics Society Conference* (2017).

[90] M. Stauffer, A. Fischer and K. Riesen, Speeding-Up Graph-Based Keyword Spotting by Quadtree Segmentations, in *International Conference on Computer Analysis of Images and Patterns*. Springer, pp. 304–315 (2017a).

[91] M. Stauffer, A. Fischer and K. Riesen, Ensembles for Graph-Based Keyword Spotting in Historical Handwritten Documents, in *International Conference on Document Analysis and Recognition*. IEEE, pp. 714–720 (2017b).

[92] M. Stauffer, A. Fischer and K. Riesen, Graph-Based Keyword Spotting in Historical Documents Using Context-Aware Hausdorff Edit Distance, in *International Workshop on Document Analysis Systems*. IEEE, pp. 49–54 (2018).

[93] K. Riesen and H. Bunke, Approximate Graph Edit Distance Computation by Means of Bipartite Graph Matching, *Image and Vision Computing* **27**, 7, pp. 950–959 (2009).

[94] A. Fischer, S. Uchida, V. Frinken, K. Riesen and H. Bunke, Improving Hausdorff Edit Distance Using Structural Node Context, in *International Workshop on Graph-Based Representations in Pattern Recognition*. Springer, pp. 148–157 (2015).

[95] A. Fischer, K. Riesen and H. Bunke, Improved Quadratic Time Approx-

imation of Graph Edit Distance by Combining Hausdorff Matching and Greedy Assignment, *Pattern Recognition Letters* **87**, pp. 55–62 (2017).

[96] V. Frinken, A. Fischer and H. Bunke, Combining Neural Networks to Improve Performance of Handwritten Keyword Spotting, in *Multiple Classifier Systems*. Springer, pp. 215–224 (2010).

[97] J. Lladós, M. Rusiñol, A. Fornés, D. Fernández and A. Dutta, On the Influence of Word Representations for Handwritten Word Spotting in Historical Documents, *International Journal of Pattern Recognition and Artificial Intelligence* **26**, 05, p. 1263002 (2012).

[98] G. Sfikas, G. Retsinas and B. Gatos, Zoning Aggregated Hypercolumns for Keyword Spotting, in *International Conference on Frontiers in Handwriting Recognition*. IEEE, pp. 283–288 (2016).

[99] A. Kolcz, J. Alspector, M. Augusteijn, R. Carlson and G. Viorel Popescu, A Line-Oriented Approach to Word Spotting in Handwritten Documents, *Pattern Analysis and Applications* **3**, 2, pp. 153–168 (2000).

[100] T. M. Rath, S. Kane, A. Lehman, E. Partridge and R. Manmatha, Indexing for a Digital Library of George Washington's Manuscripts: A Study of Word Matching Techniques, Tech. Rep., Center for Intelligent Information Retrieval, University of Massachusetts Amherst (2002).

[101] E. F. Can and P. Duygulu, A Line-Based Representation for Matching Words in Historical Manuscripts, *Pattern Recognition Letters* **32**, 8, pp. 1126–1138 (2011).

[102] V. Frinken, A. Fischer and H. Bunke, A Novel Word Spotting Algorithm Using Bidirectional Long Short-Term Memory Neural Networks, in *Workshop on Artificial Neural Networks in Pattern Recognition*. Springer, pp. 185–196 (2010a).

[103] V. Frinken, A. Fischer, H. Bunke and R. Manmatha, Adapting BLSTM Neural Network-Based Keyword Spotting Trained on Modern Data to Historical Documents, in *International Conference on Frontiers in Handwriting Recognition*. IEEE, pp. 352–357 (2010b).

[104] D. Fernández, J. Lladós and A. Fornés, Handwritten Word Spotting in Old Manuscript Images Using a Pseudo-structural Descriptor Organized in a Hash Structure, in *Iberian Conference on Pattern Recognition and Image Analysis*. Springer, pp. 628–635 (2011).

[105] L. Rothacker, S. Vajda and G. A. Fink, Bag-of-features Representations for Offline Handwriting Recognition Applied to Arabic Script, in *International Conference on Frontiers in Handwriting Recognition*. IEEE, pp. 149–154 (2012).

[106] S. Wshah, G. Kumar and V. Govindaraju, Script Independent Word Spotting in Offline Handwritten Documents Based on Hidden Markov Models, in *International Conference on Frontiers in Handwriting Recognition*. IEEE, pp. 14–19 (2012).

[107] A. Fischer, V. Frinken, H. Bunke and C. Y. Suen, Improving HMM-Based Keyword Spotting with Character Language Models, in *International Conference on Document Analysis and Recognition*. IEEE, pp. 506–510 (2013).

[108] N. R. Howe, Part-structured Inkball Models for One-shot Handwritten

Word Spotting, in *International Conference on Document Analysis and Recognition.* IEEE, pp. 582–586 (2013).

[109] A. P. Giotis, D. P. Gerogiannis and C. Nikou, Word Spotting in Handwritten Text Using Contour-Based Models, in *International Conference on Frontiers in Handwriting Recognition*, Vol. 2014-Decem. IEEE, pp. 399–404 (2014).

[110] D. Fernández-Mota, R. Manmatha, A. Fornés and J. Lladós, Sequential Word Spotting in Historical Handwritten Documents, in *International Workshop on Document Analysis Systems.* IEEE, pp. 101–105 (2014a).

[111] D. Fernández-Mota, P. Riba, A. Fornés and J. Lladós, On the Influence of Key Point Encoding for Handwritten Word Spotting, in *International Conference on Frontiers in Handwriting Recognition.* IEEE, pp. 476–481 (2014b).

[112] A. Kovalchuk, L. Wolf and N. Dershowitz, A Simple and Fast Word Spotting Method, in *International Conference on Frontiers in Handwriting Recognition.* IEEE, pp. 3–8 (2014).

[113] T. Konidaris, A. L. Kesidis and B. Gatos, A Segmentation-free Word Spotting Method for Historical Printed Documents, *Pattern Analysis and Applications* **19**, 4, pp. 963–976 (2015).

[114] N. R. Howe, Inkball Models for Character Localization and Out-of-vocabulary Word Spotting, in *International Conference on Document Analysis and Recognition.* IEEE, pp. 381–385 (2015).

[115] T. Mondal, N. Ragot, J. Y. Ramel and U. Pal, Flexible Sequence Matching Technique: An Effective Learning-free Approach for Word Spotting, *Pattern Recognition* **60**, pp. 596–612 (2016).

[116] G. Retsinas, N. Stamatopoulos, G. Louloudis, G. Sfikas and B. Gatos, Nonlinear Manifold Embedding on Keyword Spotting Using t-SNE, in *International Conference on Document Analysis and Recognition.* IEEE, pp. 487–492 (2017).

[117] R. Manmatha and T. M. Rath, Indexing of Handwritten Historical Documents - Recent Progress, in *Symposium on Document Image Understanding Technology*, pp. 77–85 (2003).

[118] M. W. Sagheer, N. Nobile, C. L. He and C. Y. Suen, A Novel Handwritten Urdu Word Spotting Based on Connected Components Analysis, in *International Conference on Pattern Recognition.* IEEE, pp. 2013–2016 (2010).

[119] A. Fischer, K. Riesen and H. Bunke, Graph Similarity Features for HMM-Based Handwriting Recognition in Historical Documents, in *International Conference on Frontiers in Handwriting Recognition.* IEEE, pp. 253–258 (2010).

[120] L. Rothacker, G. A. Fink, P. Banerjee, U. Bhattacharya and B. B. Chaudhuri, Bag-of-features HMMs for Segmentation-free Bangla Word Spotting, in *International Workshop on Multilingual OCR.* ACM, pp. 5:1–5:5 (2013).

[121] S. Wshah, G. Kumar and V. Govindaraju, Statistical Script Independent Word Spotting in Offline Handwritten Documents, *Pattern Recognition* **47**, 3, pp. 1039–1050 (2014).

[122] T. M. Rath and R. Manmatha, Features for Word Spotting in Histori-

cal Manuscripts, in *International Conference on Document Analysis and Recognition*. IEEE, pp. 218–222 (2003).

[123] A. H. Toselli, E. Vidal, V. Romero and V. Frinken, HMM Word Graph-Based Keyword Spotting in Handwritten Document Images, *Information Sciences* **370-371**, pp. 497–518 (2016).

[124] H. Cao and V. Govindaraju, Template-free Word Spotting in Low-quality Manuscripts, in *International Conference on Advances in Pattern Recognition*. World Scientific, pp. 1–5 (2007).

[125] A. Bhardwaj, D. Jose and V. Govindaraju, Script Independent Word Spotting in Multilingual Documents. in *Workshop on Cross Lingual Information Access*, pp. 48–54 (2008).

[126] Z. A. Aghbari and S. Brook, HAH Manuscripts: A Holistic Paradigm for Classifying and Retrieving Historical Arabic Handwritten Documents, *Expert Systems with Applications* **36**, 8, pp. 10942–10951 (2009).

[127] L. Huang, F. Yin, Q.-H. Chen and C.-L. Liu, Keyword Spotting in Offline Chinese Handwritten Documents Using a Statistical Model, in *International Conference on Document Analysis and Recognition*. IEEE, pp. 78–82 (2011).

[128] J. Almazán, A. Gordo, A. Fornés and E. Valveny, Handwritten Word Spotting with Corrected Attributes, in *International Conference on Computer Vision*. IEEE, pp. 1017–1024 (2013).

[129] P. Wang, V. Eglin, C. Garcia, C. Largeron, J. Lladós and A. Fornés, A Coarse-to-Fine Word Spotting Approach for Historical Handwritten Documents Based on Graph Embedding and Graph Edit Distance, in *International Conference on Pattern Recognition*. IEEE, pp. 3074–3079 (2014).

[130] K. Zagoris, I. Pratikakis and B. Gatos, Segmentation-Based Historical Handwritten Word Spotting Using Document-Specific Local Features, in *International Conference on Frontiers in Handwriting Recognition*. IEEE, pp. 9–14 (2014).

[131] D. Aldavert, M. Rusiñol, R. Toledo and J. Lladós, A Study of Bag-of-Visual-Words Representations for Handwritten Keyword Spotting, *International Journal on Document Analysis and Recognition* **18**, 3, pp. 223–234 (2015).

[132] S. Yao, Y. Wen and Y. Lu, HOG-Based Two-directional Dynamic Time Warping for Handwritten Word Spotting, in *International Conference on Document Analysis and Recognition*. IEEE, pp. 161–165 (2015).

[133] S. Sudholt and G. A. Fink, A Modified Isomap Approach to Manifold Learning in Word Spotting, in *German Conference on Pattern Recognition*. Springer, pp. 529–539 (2015).

[134] Z. Zhong, W. Pan, L. Jin, H. Mouchère and C. Viard-Gaudin, SpottingNet: Learning the Similarity of Word Images with Convolutional Neural Networks for Word Spotting in Handwritten Historical Documents, in *International Conference on Frontiers in Handwriting Recognition*. IEEE, pp. 295–300 (2016).

[135] S. Sudholt, L. Rothacker and G. A. Fink, Query-by-Online Word Spotting Revisited: Using CNNs for Cross-Domain Retrieval, in *International*

Conference on Document Analysis and Recognition. IEEE, pp. 481–486 (2017).

[136] N. Otsu, A Threshold Selection Method from Gray-Level Histograms, *Transactions on Systems, Man and Cybernetics* **9**, 1, pp. 62–66 (1979).

[137] E. Ozdemir and C. Gunduz-Demir, A Hybrid Classification Model for Digital Pathology Using Structural and Statistical Pattern Recognition, *Transactions on Medical Imaging* **32**, 2, pp. 474–483 (2013).

[138] W. Niblack, *An Introduction to Digital Image Processing.* Strandberg Publishing Company (1985).

[139] J. Sauvola and M. Pietikäinen, Adaptive Document Image Binarization, *Pattern Recognition* **33**, 2, pp. 225–236 (2000).

[140] J. A. Rodríguez-Serrano and F. Perronnin, A Model-Based Sequence Similarity with Application to Handwritten Word Spotting, *Transactions on Pattern Analysis and Machine Intelligence* **34**, 11, pp. 2108 – 2120 (2012).

[141] A. Fischer, E. Indermühle, H. Bunke, G. Viehhauser and M. Stolz, Ground Truth Creation for Handwriting Recognition in Historical Documents, in *International Workshop on Document Analysis Systems.* ACM, pp. 3–10 (2010).

[142] S. Srihari, H. Srinivasan, P. Babu and C. Bhole, Spotting Words in Handwritten Arabic Documents, in *Document Recognition and Retrieval.* International Society for Optics and Photonics, p. 606702 (2006).

[143] Y. Leydier, F. Lebourgeois and H. Emptoz, Text Search for Medieval Manuscript Images, *Pattern Recognition* **40**, 12, pp. 3552–3567 (2007).

[144] Y. Leydier, A. Ouji, F. LeBourgeois and H. Emptoz, Towards an Omnilingual Word Retrieval System for Ancient Manuscripts, *Pattern Recognition* **42**, 9, pp. 2089–2105 (2009).

[145] J. Almazán, A. Gordo, A. Fornés and E. Valveny, Efficient Exemplar Word Spotting, in *British Machine Vision Conference.* BMVA Press, pp. 67.1–67.11 (2012).

[146] X. Zhang and C. L. Tan, Segmentation-free Keyword Spotting for Handwritten Documents Based on Heat Kernel Signature, in *International Conference on Document Analysis and Recognition.* IEEE, pp. 827–831 (2013).

[147] H. Wei, G. Gao and X. Su, A Multiple Instances Approach to Improving Keyword Spotting on Historical Mongolian Document Images, in *International Conference on Document Analysis and Recognition.* IEEE, pp. 121–125 (2015).

[148] A. Vinciarelli, S. Bengio and H. Bunke, Offline Recognition of Unconstrained Handwritten Texts Using HMMs and Statistical Language Models, *Transactions on Pattern Analysis and Machine Intelligence* **26**, 6, pp. 709–720 (2004).

[149] Y. Kessentini, T. Paquet and A. Ben Hamadou, Off-line Handwritten Word Recognition Using Multi-stream Hidden Markov Models, *Pattern Recognition Letters* **31**, 1, pp. 60–70 (2010).

[150] T. M. Rath, R. Manmatha and V. Lavrenko, A Search Engine for Historical Manuscript Images, in *International Conference on Research and Development in Information Retrieval.* ACM, pp. 369–376 (2004).

[151] C.-H. Teh and R. T. Chin, On Image Analysis by the Methods of Moments, *Transactions on Pattern Analysis and Machine Intelligence* **10**, 4, pp. 496–513 (1988).

[152] J. Favata, G. Srikanta and S. N. Srihari, Handprinted Character/Digit Recognition Using a Multiple Feature/Resolution Philosophy, in *International Workshop Frontiers in Handwriting Recognition*, pp. 57–66 (1994).

[153] J. T. Favata and G. Srikantan, A Multiple Feature/Resolution Approach to Handprinted Digit and Character Recognition, *International Journal of Imaging Systems and Technology* **7**, 4, pp. 304–311 (1996).

[154] N. Dalal and B. Triggs, Histograms of Oriented Gradients for Human Detection, in *Computer Society Conference on Computer Vision and Pattern Recognition*, Vol. I. IEEE, pp. 886–893 (2005).

[155] D. Lowe, Object Recognition from Local Scale-invariant Features, in *International Conference on Computer Vision*. IEEE, pp. 1150–1157 (1999).

[156] H. A. Glucksman, Classification of Mixed-font Alphabetics by Characteristic Loci, in *First Annual IEEE Computer Conference*. IEEE, pp. 138–141 (1967).

[157] L. Wang and D.-C. He, Texture Classification Using Texture Spectrum, *Pattern Recognition* **23**, 8, pp. 905–910 (1990).

[158] D. C. He and L. Wang, Texture Unit, Texture Spectrum, and Texture Analysis, *Transactions on Geoscience and Remote Sensing* **28**, 4, pp. 509–512 (1990).

[159] S. Belongie, J. Malik and J. Puzicha, Shape Matching and Object Recognition Using Shape Contexts, *Transactions on Pattern Analysis and Machine Intelligence* **24**, 4, pp. 509–522 (2002).

[160] S. Escalera, A. Fornés, O. Pujol, P. Radeva, G. Sánchez and J. Lladós, Blurred Shape Model for Binary and Grey-level Symbol Recognition, *Pattern Recognition Letters* **30**, 15, pp. 1424–1433 (2009).

[161] V. Ferrari, F. Jurie and C. Schmid, From Images to Shape Models for Object Detection, *International Journal of Computer Vision* **87**, pp. 284–303 (2009).

[162] J. Sun, M. Ovsjanikov and L. Guibas, A Concise and Provably Informative Multi-scale Signature Based on Heat Diffusion, in *Eurographics Symposium on Geometry Processing*, Vol. 28. Wiley, pp. 1383–1392 (2009).

[163] G. Csurka, C. Dance, L. Fan, J. Willamowski and Cedric Bray, Visual Categorization with Bag of Keypoints, in *International Workshop on Statistical Learning in Computer Vision*, pp. 1–22 (2004).

[164] F. Perronnin and C. Dance, Fisher Kernels on Visual Vocabularies for Image Categorization, in *Computer Society Conference on Computer Vision and Pattern Recognition*. IEEE (2007).

[165] T. Jaakkola and D. Haussler, Exploiting Generative Models in Discriminative Classifiers, in *Advances in Neural Information Processing Systems*. MIT Press, pp. 487–493 (1998).

[166] J. Sánchez, F. Perronnin, T. Mensink and J. Verbeek, Image Classification with the Fisher Vector: Theory and Practice, *International Journal of Computer Vision* **105**, 3, pp. 222–245 (2013).

[167] K. Pearson, On Lines and Planes of Closest Fit to Systems of Points in Space, *The London, Edinburgh, and Dublin Philosophical Magazine and Journal of Science* **2**, 11, pp. 559–572 (1901).

[168] J. B. Tenenbaum, A Global Geometric Framework for Nonlinear Dimensionality Reduction, *Science* **290**, 5500, pp. 2319–2323 (2000).

[169] L. V. D. Maaten and G. Hinton, Visualizing Data using t-SNE, *Journal of Machine Learning Research* **620**, 1, pp. 267–84 (2008).

[170] L. Rothacker, S. Sudholt, E. Rusakov, M. Kasperidus and G. A. Fink, Word Hypotheses for Segmentation-free Word Spotting in Historic Document Images, in *International Conference on Document Analysis and Recognition.* IEEE, pp. 1174–1179 (2017).

[171] V. Dovgalecs, A. Burnett, P. Tranouez, S. Nicolas and L. Heutte, Spot It! Finding Words and Patterns in Historical Documents, in *International Conference on Document Analysis and Recognition.* IEEE, pp. 1039–1043 (2013).

[172] K. Riesen, D. Brodić, Z. N. Milivojević and Č. A. Maluckov, Graph Based Keyword Spotting in Medieval Slavic Documents – A Project Outline, in *International Conference on Cultural Heritage.* Springer, pp. 724–731 (2014).

[173] K. Riesen, M. Ferrer, A. Fischer and H. Bunke, Approximation of Graph Edit Distance in Quadratic Time, in *International Workshop on Graph-Based Representations in Pattern Recognition*, Vol. 4538. Springer, pp. 3–12 (2015).

[174] D. Fernández-Mota, J. Lladós and A. Fornés, A Graph-Based Approach for Segmenting Touching Lines in Historical Handwritten Documents, *International Journal on Document Analysis and Recognition* **17**, 3, pp. 293–312 (2014).

[175] L. Rothacker and G. A. Fink, Segmentation-free Query-by-string Word Spotting with Bag-of-Features HMMs, in *International Conference on Document Analysis and Recognition.* IEEE, pp. 661–665 (2015).

[176] A. Graves, M. Liwicki, S. Fernández, R. Bertolami, H. Bunke and J. Schmidhuber, A Novel Connectionist System for Improved Unconstrained Handwriting Recognition, *Transactions on Pattern Analysis and Machine Intelligence* **31**, 5, pp. 1–14 (2009).

[177] S. Deerwester, S. T. Dumais, G. W. Furnas, T. K. Landauer and R. Harshman, Indexing by Latent Semantic Analysis, *Journal of the American Society for Information Science* **41**, 6, pp. 391–407 (1990).

[178] S. Ghosh and E. Valveny, R-PHOC: Segmentation-free Word Spotting Using CNN, in *International Conference on Document Analysis and Recognition.* IEEE, pp. 801–806 (2017).

[179] P. Riba, A. Fischer, J. Lladós and A. Fornés, Learning Graph Distances with Message Passing Neural Networks, in *International Conference on Pattern Recognition.* IEEE (2018).

[180] Z. Guo and R. W. Hall, Parallel Thinning with Two-subiteration Algorithms, *Communications of the ACM* **32**, 3, pp. 359–373 (1989).

[181] A. Fischer, M. Baechler, A. Garz, M. Liwicki and R. Ingold, A Combined System for Text Line Extraction and Handwriting Recognition in Historical

Documents, in *International Workshop on Document Analysis Systems.* IEEE, pp. 71–75 (2014).

[182] R. van den Boomgaard and R. van Balen, Methods for Fast Morphological Image Transforms Using Bitmapped Binary Images, *CVGIP: Graphical Models and Image Processing* **54**, 3, pp. 252–258 (1992).

[183] K. Riesen, *Structural Pattern Recognition with Graph Edit Distance*, Advances in Computer Vision and Pattern Recognition. Springer (2015).

[184] P. Mahé, N. Ueda, T. Akutsu, J.-L. Perret and J.-P. Vert, Graph Kernels for Molecular Structure-Activity Relationship Analysis with Support Vector Machines. *Journal of Chemical Information and Modeling* **45**, 4, pp. 939–951 (2005).

[185] A. Schenker, A. Kandel, H. Bunke and M. Last, *Graph-Theoretic Techniques for Web Content Mining, Series in Machine Perception and Artificial Intelligence*, Vol. 62. World Scientific (2005).

[186] K. M. Borgwardt, H.-P. Kriegel, S. V. N. Vishwanathan and N. N. Schraudolph, Graph Kernels for Disease Outcome Prediction from Protein-Protein Interaction Networks. in *Pacific Symposium on Biocomputing*. World Scientific, pp. 4–15 (2007).

[187] P. J. Dickinson, H. Bunke, A. Dadej and M. Kraetzl, Matching Graphs with Unique Node Labels, *Pattern Analysis and Applications* **7**, 3, pp. 243–254 (2004).

[188] D. Conte, P. Foggia, C. Sansone and M. Vento, Graph Matching Applications in Pattern Recognition and Image Processing, in *International Conference on Image Processing*, Vol. 3. IEEE, pp. 21–24 (2003).

[189] P. Foggia, G. Percannella and M. Vento, Graph Matching and Learning in Pattern Recognition in the Last 10 Years, *International Journal of Pattern Recognition and Artificial Intelligence* **28**, 01, p. 1450001 (2014).

[190] J. B. Kruskal, On the Shortest Spanning Subtree of a Graph and the Traveling Salesman Problem, *Proceedings of the American Mathematical Society* **7**, 1, pp. 48–50 (1956).

[191] T. Pavlidis, A Minimum Storage Boundary Tracing Algorithm and Its Application to Automatic Inspection, *Transactions on Systems, Man and Cybernetics* **8**, 1, pp. 66–69 (1978).

[192] G. Monagan and M. Roosli, Appropriate Base Representation Using a Run Graph, in *International Conference on Document Analysis and Recognition*. IEEE, pp. 623–626 (1993).

[193] L. Boatto, V. Consorti, M. Del Buono, S. Di Zenzo, V. Eramo, A. Esposito, F. Melcarne, M. Meucci, A. Morelli, M. Mosciatti, S. Scarci and M. Tucci, An Interpretation System for Land Register Maps, *Computer* **25**, 7, pp. 25–33 (1992).

[194] M. Burge and W. G. Kropatsch, A Minimal Line Property Preserving Representation of Line Images, *Computing* **62**, 4, pp. 355–368 (1999).

[195] H. Bunke and K. Riesen, Recent Advances in Graph-Based Pattern Recognition with Applications in Document Analysis, *Pattern Recognition* **44**, 5, pp. 1057–1067 (2011).

[196] K. Riesen, A. Fischer and H. Bunke, Combining Bipartite Graph Matching

and Beam Search for Graph Edit Distance Approximation, in *Workshop on Artificial Neural Networks in Pattern Recognition*. Springer, pp. 117–128 (2014).

[197] P. Foggia and M. Vento, Graph Embedding for Pattern Recognition, in *Recognizing Patterns in Signals Speech Images and Videos*. Springer, pp. 75–82 (2010).

[198] K. Riesen and H. Bunke, *Graph Classification and Clustering Based on Vector Space Embedding*, Series in Machine Perception and Artificial Intelligence. World Scientific (2010).

[199] D. Conte, J.-Y. Ramel, N. Sidère, M. M. Luqman, B. Gaüzère, J. Gibert, L. Brun and M. Vento, A Comparison of Explicit and Implicit Graph Embedding Methods for Pattern Recognition, in *International Workshop on Graph-Based Representations in Pattern Recognition*. Springer, pp. 81–90 (2013).

[200] T. Gärtner, J. W. Lloyd and P. A. Flach, Kernels for Structured Data, in *International Conference on Inductive Logic Programming*. Springer, pp. 66–83 (2003).

[201] M. Neuhaus and H. Bunke, *Bridging the Gap between Graph Edit Distance and Kernel Machines*. World Scientific (2007).

[202] N. M. Kriege and C. Morris, Recent Advances in Kernel-Based Graph Classification, in *Joint European Conference on Machine Learning and Knowledge Discovery in Databases*. Springer, pp. 388–392 (2017).

[203] D. Conte, P. Foggia, C. Sansone and M. Vento, Thirty Years of Graph Matching in Pattern Recognition, *International Journal of Pattern Recognition and Artificial Intelligence* **18**, 03, pp. 265–298 (2004).

[204] M. Vento, A Long Trip in the Charming World of Graphs for Pattern Recognition, *Pattern Recognition* **48**, 2, pp. 291–301 (2015).

[205] M. R. Garey and D. S. Johnson, *Computers and Intractability: A Guide to the Theory of NP-Completeness*. W. H. Freeman and Company (1979).

[206] L. Babai, Graph Isomorphism in Quasipolynomial Time, in *Symposium on Theory of Computing*. ACM, pp. 684–697 (2016).

[207] J. R. Ullmann, An Algorithm for Subgraph Isomorphism, *Journal of the ACM* **23**, 1, pp. 31–42 (1976).

[208] L. P. Cordella, P. Foggia, C. Sansone and M. Vento, A (Sub)graph Isomorphism Algorithm for Matching Large Graphs, *Transactions on Pattern Analysis and Machine Intelligence* **26**, 10, pp. 1367–1372 (2004).

[209] L. Cordella, P. Foggia, C. Sansone and M. Vento, Performance Evaluation of the VF Graph-Matching Algorithm, in *International Conference on Image Analysis and Processing*. IEEE, pp. 3–8 (1999).

[210] J. Larrosa and G. Valiente, Constraint Satisfaction Algorithms for Graph Pattern Matching, *Mathematical Structures in Computer Science* **12**, 4, pp. 403–422 (2002).

[211] S. Zampelli, Y. Deville and C. Solnon, Solving Subgraph Isomorphism Problems with Constraint Programming, *Constraints* **15**, 3, pp. 327–353 (2010).

[212] J. R. Ullmann and J. R., Bit-vector Algorithms for Binary Constraint Satis-

faction and Subgraph Isomorphism, *Journal of Experimental Algorithmics* **15**, p. 1.1 (2010).

[213] C. Solnon, All Different-Based Filtering for Subgraph Isomorphism, *Artificial Intelligence* **174**, 12-13, pp. 850–864 (2010).

[214] N. Dahm, H. Bunke, T. Caelli and Y. Gao, Efficient Subgraph Matching Using Topological Node Feature Constraints, *Pattern Recognition* **48**, 2, pp. 317–330 (2015).

[215] V. Carletti, P. Foggia, A. Saggese and M. Vento, Introducing VF3: A New Algorithm for Subgraph Isomorphism, in *International Workshop on Graph-Based Representations in Pattern Recognition*. Springer, pp. 128–139 (2017).

[216] B. D. McKay, Practical Graph Isomorphism, *Congressus Numerantium* **30**, pp. 45–87 (1981).

[217] B. D. McKay and A. Piperno, Practical Graph Isomorphism, II, *Journal of Symbolic Computation* **60**, pp. 94–112 (2014).

[218] M. Gori, M. Maggini and L. Sarti, Exact and Approximate Graph Matching Using Random Walks, *Transactions on Pattern Analysis and Machine Intelligence* **27**, 7, pp. 1100–1111 (2005).

[219] M. Cho, J. Lee and K. M. Lee, Reweighted Random Walks for Graph Matching, in *European Conference on Computer Vision*. Springer, pp. 492–505 (2010).

[220] B. T. Messmer and H. Bunke, A Decision Tree Approach to Graph and Subgraph Isomorphism Detection, *Pattern Recognition* **32**, 12, pp. 1979–1998 (1999).

[221] M. Weber, M. Liwicki and A. Dengel, Faster Subgraph Isomorphism Detection by Well-founded Total Order Indexing, *Pattern Recognition Letters* **33**, 15, pp. 2011–2019 (2012).

[222] K. Riesen, X. Jiang and H. Bunke, Exact and Inexact Graph Matching: Methodology and Applications, in *Managing and Mining Graph Data*, Vol. 40. Springer, pp. 217–247 (2010).

[223] H. Bunke, X. Jiang and A. Kandel, On the Minimum Common Supergraph of Two Graphs, *Computing* **65**, 1, pp. 13–25 (2000).

[224] G. Levi, A Note on the Derivation of Maximal Common Subgraphs of Two Directed or Undirected Graphs, *Calcolo* **9**, 4, pp. 341–352 (1973).

[225] J. J. McGregor, Backtrack Search Algorithms and the Maximal Common Subgraph Problem, *Software: Practice and Experience* **12**, 1, pp. 23–34 (1982).

[226] M. C. Boeres, C. C. Ribeiro and I. Bloch, A Randomized Heuristic for Scene Recognition by Graph Matching, in *Experimental and Efficient Algorithms*, Vol. 3059. Springer, pp. 100–113 (2004).

[227] M. Neuhaus, K. Riesen and H. Bunke, Fast Suboptimal Algorithms for the Computation of Graph Edit Distance, in *International Workshop on Structural, Syntactic, and Statistical Pattern Recognition*. Springer, pp. 163–172 (2006).

[228] F. Serratosa, Fast Computation of Bipartite Graph Matching, *Pattern Recognition Letters* **45**, 1, pp. 244–250 (2014).

[229] J. T. Vogelstein, J. M. Conroy, V. Lyzinski, L. J. Podrazik, S. G. Kratzer, E. T. Harley, D. E. Fishkind, R. J. Vogelstein and C. E. Priebe, Fast Approximate Quadratic Programming for Graph Matching, *PLoS ONE* **10**, 4, p. e0121002 (2015).

[230] F. Serratosa, Speeding up Fast Bipartite Graph Matching Through a New Cost Matrix, *International Journal of Pattern Recognition and Artificial Intelligence* **29**, 02, p. 1550010 (2015).

[231] V. Lyzinski, D. E. Fishkind, M. Fiori, J. T. Vogelstein, C. E. Priebe and G. Sapiro, Graph Matching: Relax at Your Own Risk, *Transactions on Pattern Analysis and Machine Intelligence* **38**, 1, pp. 60–73 (2016).

[232] T. Wang, H. Ling, C. Lang and S. Feng, Graph Matching with Adaptive and Branching Path Following, *Transactions on Pattern Analysis and Machine Intelligence* , 99, pp. 1–1 (2017).

[233] A. Sanfeliu, R. Alquézar, J. Andrade, J. Climent, F. Serratosa and J. Vergés, Graph-Based Representations and Techniques for Image Processing and Image Analysis, *Pattern Recognition* **35**, 3, pp. 639–650 (2002).

[234] F. Serratosa, R. Alquézar and A. Sanfeliu, Function-described Graphs for Modelling Objects Represented by Sets of Attributed Graphs, *Pattern Recognition* **36**, 3, pp. 781–798 (2003).

[235] D. J. Cook, N. Manocha and L. B. Holder, Using a Graph-Based Data Mining System to Perform Web Search, *International Journal of Pattern Recognition and Artificial Intelligence* **17**, 5, pp. 705–720 (2003).

[236] S. C. Sorlin S., Reactive Tabu Search for Measuring Graph Similarity, in *International Workshop on Graph-Based Representations in Pattern Recognition.* Springer, pp. 172–182 (2005).

[237] K. Adamczewski, Y. Suh and K. M. Lee, Discrete Tabu Search for Graph Matching, in *International Conference on Computer Vision.* IEEE, pp. 109–117 (2015).

[238] A. Massaro and M. Pelillo, Matching Graphs by Pivoting, *Pattern Recognition Letters* **24**, 8, pp. 1099–1106 (2003).

[239] M. Zaslavskiy, F. Bach and J. Vert, A Path Following Algorithm for the Graph Matching Problem, *Transactions on Pattern Analysis and Machine Intelligence* **31**, 12, pp. 2227–2242 (2009).

[240] D. Justice and A. Hero, A Binary Linear Programming Formulation of the Graph Edit Distance, *Transactions on Pattern Analysis and Machine Intelligence* **28**, 8, pp. 1200–1214 (2006).

[241] A. Solé-Ribalta and F. Serratosa, Models and Algorithms for Computing the Common Labelling of a Set of Attributed Graphs, *Computer Vision and Image Understanding* **115**, 7, pp. 929–945 (2011).

[242] S. Umeyama, An Eigendecomposition Approach to Weighted Graph Matching Problems, *Transactions on Pattern Analysis and Machine Intelligence* **10**, 5, pp. 695–703 (1988).

[243] B. Luo, R. C. Wilson and E. R. Hancock, Spectral Feature Vectors for Graph Clustering, in *International Workshop on Structural, Syntactic, and Statistical Pattern Recognition.* Springer, pp. 83–93 (2002).

[244] T. Caelli and S. Kosinov, An Eigenspace Projection Clustering Method for

Inexact Graph Matching, *Transactions on Pattern Analysis and Machine Intelligence* **26**, 4, pp. 515–519 (2004).

[245] A. Shokoufandeh, D. Macrini, S. Dickinson, K. Siddiqi and S. Zucker, Indexing Hierarchical Structures Using Graph Spectra, *Transactions on Pattern Analysis and Machine Intelligence* **27**, 7, pp. 1125–1140 (2005).

[246] R. C. Wilson, E. R. Hancock and B. Luo, Pattern Vectors from Algebraic Graph Theory, *Transactions on Pattern Analysis and Machine Intelligence* **27**, 7, pp. 1112–1124 (2005).

[247] A. Robles-Kelly and E. R. Hancock, Graph Edit Distance from Spectral Seriation, *Transactions on Pattern Analysis and Machine Intelligence* **27**, 3, pp. 365–378 (2005).

[248] M. Yin, J. Gao and Z. Lin, Laplacian Regularized Low-Rank Representation and Its Applications, *Transactions on Pattern Analysis and Machine Intelligence* **38**, 3, pp. 504–517 (2016).

[249] T. Gärtner, P. Flach and S. Wrobel, On Graph Kernels: Hardness Results and Efficient Alternatives, in *Learning Theory and Kernel Machines*. Springer, pp. 129–143 (2003).

[250] K. M. Borgwardt, C. S. Ong, S. Schönauer, S. V. N. Vishwanathan, A. J. Smola and H. P. Kriegel, Protein Function Prediction via Graph Kernels, *Bioinformatics* **21**, 1, pp. 47–56 (2005).

[251] L. Ralaivola, S. J. Swamidass, H. Saigo and P. Baldi, Graph Kernels for Chemical Informatics, *Neural Networks* **18**, 8, pp. 1093–1110 (2005).

[252] K. M. Borgwardt and H. P. Kriegel, Shortest-path Kernels on Graphs, in *International Conference on Data Mining*. IEEE, pp. 74–81 (2005).

[253] M. Sugiyama and K. Borgwardt, Halting in Random Walk Kernels, in *Advances in Neural Information Processing Systems*, Section 2. MIT Press, pp. 1639–1647 (2015).

[254] D. Haussler, Convolution Kernels on Discrete Structures UCSC-CRL-99-10, Tech. Rep., Jack Baskin School of Engineering (1999).

[255] C. Watkins, Dynamic Alignment Kernels, *Advances in Large Margin Classifiers* , January, pp. 39–50 (1999).

[256] K. M. Borgwardt, T. Petri, S. V. N. Vishwanathan and H.-P. Kriegel, An Efficient Sampling Scheme for Comparison of Large Graphs, in *International Workshop on Mining and Learning with Graphs*, pp. 1–3 (2007).

[257] L. Rossi, A. Torsello and E. R. Hancock, A Continuous-time quantum Walk Kernel for Unattributed Graphs, in *International Workshop on Graph-Based Representations in Pattern Recognition*. Springer, pp. 101–110 (2013).

[258] L. Bai, L. Rossi, A. Torsello and E. R. Hancock, A Quantum Jensen-Shannon Graph Kernel for Unattributed Graphs, *Pattern Recognition* **48**, 2, pp. 1–12 (2014).

[259] L. Bai, L. Rossi, L. Cui, Z. Zhang, P. Ren, X. Bai and E. Hancock, Quantum Kernels for Unattributed Graphs Using Discrete-time Quantum Walks, *Pattern Recognition Letters* **87**, pp. 96–103 (2017).

[260] J.-P. Vert and M. Kanehisa, Graph-driven Features Extraction from Microarray Data Using Diffusion Kernels and Kernel CCA, in *International*

Conference on Neural Information Processing Systems. MIT Press, pp. 1449–1405 (2002).

[261] J. Lafferty and R. I. Kondor, Diffusion Kernels on Graphs and Other Discrete Input Spaces, in *International Conference on Machine Learning.* ACM, pp. 315–322 (2002).

[262] J. Lafferty, J. Lafferty, G. Lebanon and G. Lebanon, Information Diffusion Kernels, in *Advances in Neural Information Processing Systems.* MIT Press, pp. 391–398 (2003).

[263] H. Bunke and G. Allermann, Inexact Graph Matching for Structural Pattern Recognition, *Pattern Recognition Letters* **1**, 4, pp. 245–253 (1983).

[264] A. Sanfeliu and K. S. Fu, A Distance Measure Between Attributed Relational Graphs for Pattern Recognition, *Transactions on Systems, Man and Cybernetics* **13**, 3, pp. 353–362 (1983).

[265] T. C. Koopmans and M. Beckmann, Assignment Problems and the Location of Economic Activities, *Econometrica* **25**, 1, p. 53 (1957).

[266] V. I. Levenshtein, Binary Codes Capable of Correcting Deletions, Insertions, and Reversals, *Soviet Physics Doklady* **10**, 8, pp. 707–710 (1966).

[267] S. M. Selkow, The Tree-to-tree Editing Problem, *Information Processing Letters* **6**, 6, pp. 184–186 (1977).

[268] W. H. Tsai and K. S. Fu, Error-Correcting Isomorphisms of Attributed Relational Graphs for Pattern Analysis, *Transactions on Systems, Man and Cybernetics* **9**, 12, pp. 757–768 (1979).

[269] W. H. Tsai and K. S. Fu, Subgraph Error-Correcting Isomorphisms for Syntactic Pattern Recognition, *Transactions on Systems, Man and Cybernetics* **SMC-13**, 1, pp. 48–62 (1983).

[270] M. A. Eshera and K. S. Fu, A Graph Distance Measure for Image Analysis, *Transactions on Systems, Man and Cybernetics* **SMC-14**, 3, pp. 398–408 (1984).

[271] R. Ambauen, S. Fischer and H. Bunke, Graph Edit Distance with Node Splitting and Merging, and Its Application to Diatom Identification, in *International Workshop on Graph-Based Representations in Pattern Recognition.* Springer, pp. 95–106 (2003).

[272] K. Riesen, A. Fischer and H. Bunke, Improved Graph Edit Distance Approximation with Simulated Annealing, in *International Workshop on Graph-Based Representations in Pattern Recognition.* Springer, pp. 222–231 (2017).

[273] R. Burkard, M. Dell'Amico and S. Martello, *Assignment Problems.* Society for Industrial and Applied Mathematics (2009).

[274] J. Lerouge, Z. Abu-Aisheh, R. Raveaux, P. Héroux and S. Adam, New Binary Linear Programming Formulation to Compute the Graph Edit Distance, *Pattern Recognition* **72**, pp. 254–265 (2017).

[275] S. Bougleux, L. Brun, V. Carletti, P. Foggia, B. Gaüzère and M. Vento, Graph Edit Distance as a Quadratic Assignment Problem, *Pattern Recognition Letters* **87**, pp. 38–46 (2017).

[276] Z. Abu-Aisheh, R. Raveaux and J.-Y. Ramel, Anytime Graph Matching, *Pattern Recognition Letters* **84**, pp. 215–224 (2016).

[277] D. B. Blumenthal and J. Gamper, Improved Lower Bounds for Graph Edit Distance, *Transactions on Knowledge and Data Engineering* **30**, 3, pp. 503–516 (2018).

[278] K. Riesen, A. Fischer and H. Bunke, Estimating Graph Edit Distance Using Lower and Upper Bounds of Bipartite Approximations, *International Journal of Pattern Recognition and Artificial Intelligence* **29**, 02, p. 1550011 (2015).

[279] K. Riesen and H. Bunke, Improving Bipartite Graph Edit Distance Approximation Using Various Search Strategies, *Pattern Recognition* **48**, 4, pp. 1349–1363 (2015).

[280] D. B. Blumenthal and J. Gamper, Correcting and Speeding-up Bounds for Non-uniform Graph Edit Distance, in *International Conference on Data Engineering*. IEEE, pp. 131–134 (2017).

[281] S. Fankhauser, K. Riesen and H. Bunke, Speeding Up Graph Edit Distance Computation through Fast Bipartite Matching, in *International Workshop on Graph-Based Representations in Pattern Recognition*. Springer, pp. 102–111 (2011).

[282] B. Gaüzère, S. Bougleux, K. Riesen and L. Brun, Approximate Graph Edit Distance Guided by Bipartite Matching of Bags of Walks, in *International Workshop on Structural, Syntactic, and Statistical Pattern Recognition*. Springer, pp. 73–82 (2014).

[283] K. Riesen and H. Bunke, Improving Approximate Graph Edit Distance by Means of a Greedy Swap Strategy, in *International Conference on Image and Signal Processing*. Springer, pp. 314–321 (2014).

[284] K. Riesen, H. Bunke and A. Fischer, Improving Graph Edit Distance Approximation by Centrality Measures, in *International Conference on Pattern Recognition*. IEEE, pp. 3910–3914 (2014a).

[285] K. Riesen, A. Fischer and H. Bunke, Improving Approximate Graph Edit Distance Using Genetic Algorithms, in *International Workshop on Structural, Syntactic, and Statistical Pattern Recognition*. Springer, pp. 63–72 (2014b).

[286] V. Carletti, B. Gaüzère, L. Brun and M. Vento, Approximate Graph Edit Distance Computation Combining Bipartite Matching and Exact Neighborhood Substructure Distance, in *International Workshop on Graph-Based Representations in Pattern Recognition*. Springer, pp. 188–197 (2015).

[287] F. Serratosa and X. Cortés, Graph Edit Distance: Moving from Global to Local Structure to Solve the Graph-matching Problem, *Pattern Recognition Letters* **65**, pp. 204–210 (2015).

[288] M. Ferrer, F. Serratosa and K. Riesen, Improving Bipartite Graph Matching by Assessing the Assignment Confidence, *Pattern Recognition Letters* **65**, pp. 29–36 (2015a).

[289] M. Ferrer, F. Serratosa and K. Riesen, A First Step Towards Exact Graph Edit Distance Using Bipartite Graph Matching, in *International Workshop on Graph-Based Representations in Pattern Recognition*. Springer, pp. 77–86 (2015b).

[290] M. Ferrer, F. Serratosa and K. Riesen, Learning Heuristics to Reduce the Overestimation of Bipartite Graph Edit Distance Approximation, in *International Workshop on Machine Learning and Data Mining in Pattern Recognition.* Springer, pp. 17–31 (2015c).

[291] X. Cortés, F. Serratosa and K. Riesen, On the Relevance of Local Neighbourhoods for Greedy Graph Edit Distance, in *International Workshop on Structural, Syntactic, and Statistical Pattern Recognition.* Springer, pp. 121–131 (2016).

[292] K. Riesen and M. Ferrer, Predicting the Correctness of Node Assignments in Bipartite Graph Matching, *Pattern Recognition Letters* **69**, pp. 8–14 (2016).

[293] J. Munkres, Algorithms for the Assignment and Transportation Problems, *Journal of the Society for Industrial and Applied Mathematics* **5**, 1, pp. 32–38 (1957).

[294] R. Jonker and A. Volgenant, A Shortest Augmenting Path Algorithm for Dense and Sparse Linear Assignment Problems, *Computing* **38**, 4, pp. 325–340 (1987).

[295] D. P. Huttenlocher, G. A. Klanderman and W. J. Rucklidge, Comparing Images Using the Hausdorff Distance, *Transactions on Pattern Analysis and Machine Intelligence* **15**, 9, pp. 850–863 (1993).

[296] D. Huttenlocher, W. Rucklidge and G. Klanderman, Comparing Images Using the Hausdorff Distance Under Translation, in *Computer Society Conference on Computer Vision and Pattern Recognition.* IEEE, pp. 654–656 (1992).

[297] X. Shu and X. J. Wu, A Novel Contour Descriptor for 2D Shape Matching and Its Application to Image Retrieval, *Image and Vision Computing* **29**, 4, pp. 286–294 (2011).

[298] Y. Rubner, The Earth Mover's Distance as a Metric for Image Retrieval, *International Journal of Computer Vision* **40**, 2, pp. 99–121 (2000).

[299] L. I. Kuncheva, *Combining Pattern Classifiers.* Wiley (2004).

[300] H. Sakoe and S. Chiba, Dynamic Programming Algorithm Optimization for Spoken Word Recognition, *Transactions on Acoustics, Speech, and Signal Processing* **26**, 1, pp. 43–49 (1978).

[301] A. Fischer, R. Plamondon, Y. Savaria, K. Riesen and H. Bunke, A Hausdorff Heuristic for Efficient Computation of Graph Edit Distance, in *International Workshop on Structural, Syntactic, and Statistical Pattern Recognition*, Vol. 8621. Springer, pp. 83–92 (2014).

[302] W. G. Kropatsch, Building Irregular Pyramids by Dual-graph Contraction, *IEE Proceedings - Vision, Image and Signal Processing* **142**, 6, pp. 366–374 (1994).

[303] L. Brun and W. G. Kropatsch, Introduction to Combinatorial Pyramids, in *Digital and Image Geometry.* Springer, pp. 108–128 (2001).

[304] W. G. Kropatsch, Y. Haxhimusa, Z. Pizlo and G. Langs, Vision Pyramids that Do Not Grow Too High, *Pattern Recognition Letters* **26**, 3, pp. 319–337 (2005).

[305] M. Cerman, R. Gonzalez-Diaz and W. G. Kropatsch, LBP and Irregular Graph Pyramids, in *International Conference on Computer Analysis of Images and Patterns*. Springer, pp. 687–699 (2015).

[306] A. Dutta, P. Riba, J. Lladós and A. Fornés, Pyramidal Stochastic Graphlet Embedding for Document Pattern Classification, in *International Conference on Document Analysis and Recognition*. IEEE, pp. 33–38 (2017).

[307] M. M. Bronstein, J. Bruna, Y. Lecun, A. Szlam and P. Vandergheynst, Geometric Deep Learning: Going Beyond Euclidean Data, *Signal Processing Magazine* **34**, 4, pp. 18–42 (2017).

[308] P. Riba, A. Dutta, J. Lladós and A. Fornés, Graph-Based Deep Learning for Graphics Classification, in *International Conference on Document Analysis and Recognition*. IEEE, pp. 29–30 (2017).

[309] N. M. Kriege, M. Fey, D. Fisseler, P. Mutzel and F. Weichert, Recognizing Cuneiform Signs Using Graph Based Methods, in *International Workshop on Cost-Sensitive Learning*, pp. 88:31–44 (2018).

Index

Printed in the United States
By Bookmasters